ADVENTURES
OF A MATHEMATICIAN

S. M. ULAM

Adventures of a Mathematician

ILLUSTRATED WITH PHOTOGRAPHS

CHARLES SCRIBNER'S SONS · New York

Copyright © 1976 S. M. Ulam

Library of Congress Cataloging in Publication Data
Ulam, Stanislaw M
    Adventures of a mathematician.

    Bibliography: p.
    Includes index.
    1. Ulam, Stanislaw M.   I. Title.
QA29.U4A33   510'.92'4 [B]   75-20133
ISBN 0-684-14391-7

1 3 5 7 9 11 13 15 17 19   v|c   20 18 16 14 12 10 8 6 4 2

Printed in the United States of America

In memory of my parents

ACKNOWLEDGMENTS

This book could have been written without the help of my wife, Françoise, but it would have been merely a chaotic assemblage of items. That it may present some coherent features is the result of her intervention and collaboration. She managed to decrease substantially the entropy of this collection of reminiscences through several years' intelligent and systematic work. Thanks are also due to Gian-Carlo Rota, for our numerous conversations on some of the topics of this book; to Mrs. Emilia Mycielska, for her research on my deceased Polish colleagues; and to Mrs. Jane Richtmyer for going over some of the rougher spots of the text.

Grateful acknowledgment for permission to reproduce photographs is made to the following:
The Society of Fellows, Harvard (Harvard Junior Fellows, 1938)
Los Alamos Scientific Laboratory (all photographs so credited in the captions)
Harold Agnew (Enrico Fermi in the 1940s)
Nicholas Metropolis (Von Neumann, Feynman, and Ulam at Bandelier Lodge)
The Viking Press (George Gamow's cartoon of the "super" committee). From *My World Line* by George Gamow; copyright © 1970 by George Gamow.
Lloyd Shearer (Stan and Françoise Ulam at home, 1964)

All pictures not otherwise credited are the property of the author.

CONTENTS

CONTENTS

ILLUSTRATIONS

(following page 138)

ADVENTURES
OF A MATHEMATICIAN

Prologue

At dusk the plane from Washington to Albuquerque approached the Sandia Mountain range at the foot of which nestles the city of Albuquerque. Some ten minutes before the landing, the lights of the city of Santa Fe became visible in the distance. On the Western horizon loomed the mysterious mass of the volcanic Jemez Mountains. It was perhaps the hundredth time I was returning from Washington, New York, or California, where Los Alamos affairs or some other government or academic business took me almost every month.

My thoughts traveled back to my first arrival in New Mexico in January of 1944. I was a young professor at the University of Wisconsin and had been called to participate in a project, the exact nature of which could not be divulged at the time. All I was told was how to get to the Los Alamos area—a train station named Lamy near Santa Fe.

If someone had prophesied some forty-five years ago

(3)

that I, a young "pure" mathematician from Lwów, Poland, would spend a good part of my adult life in New Mexico—a state whose name and existence I was not even aware of when I lived in Europe—I would have dismissed the idea as inconceivable.

I found myself recollecting my childhood in Poland, my studies, my preoccupation with mathematics even at an early age, and how my interest in physics led me to enlarge my scientific curiosity, which in turn—by a series of accidents and chance—led to a call to join the Los Alamos Project. The nature of the work there I only vaguely guessed when my friend John von Neumann asked me to join him and other physicists at a strange place. "West of the Rio Grande," was all he could tell me when I met him between trains at Union Station in Chicago.

The plane landed at Albuquerque. I took my bags, walked a hundred yards across a parking area, and climbed into the small plane that commuted several times a day between Albuquerque and a single runway at an altitude of 7300 feet on the Los Alamos mesa.

Von Neumann, one of the greatest mathematicians of the first half of the twentieth century, was the person who had been responsible for my coming to this country in 1936. We had corresponded since 1934 about some abstruse questions of pure mathematics. It was in this field that I early made a name for myself; von Neumann, working in similar areas, invited me to visit the newly established Institute for Advanced Studies in Princeton—a place well known to the general public because one of its first professors was Albert Einstein. Von Neumann himself was one of the youngest professors at Princeton. He was already famous for his work in the foundations of mathematics and logic. Years later, he was to become one of the pioneers in the development of electronic computers.

At one time I had undertaken to write a book on von Neumann's scientific life. In trying to plan it, I thought of how I, along with many others, had been influenced by him;

and how this man, and some others I knew, working in the purely abstract realm of mathematics and theoretical physics had changed aspects of the world as we now know it.

Memories of my own work in science, of my studies and early research, of the endless hours spent in cafés in my home town discussing mathematics with fellow mathematicians, of my coming to the United States, lecturing at Princeton and Harvard, became interwoven in an inextricable way with recollections of von Neumann's life and later events.

When I started to organize my thoughts, I realized that up to that time—it was about 1966, I think—there existed few descriptions of the unusual climate in which the birth of the atomic age took place. Official histories do not give the real motivations or go into the inner feelings, doubts, convictions, determination, and hopes of the individuals who for over two years lived under unusual conditions. A set of flat pictures, they give at best only the essential facts.

Thinking of all this in the little plane from Albuquerque to Los Alamos, I remembered how Jules Verne and H. G. Wells had influenced me in my childhood in the books I read in Polish translation. Even in my boyish dreams, I did not imagine that some day I would take part in equally fantastic undertakings.

The result of all these reflections was that instead of writing a life of von Neumann, I have undertaken to describe my personal history, as well as what I know of a number of other scientists who also became involved in the great technological achievements of this age.

As I have already mentioned, I began as a pure mathematician. In Los Alamos I met physicists and other "natural" scientists, and consorted mainly, if not exclusively, with theoreticians. It is still an unending source of surprise for me to see how a few scribbles on a blackboard or on a sheet of paper could change the course of human affairs.

I became involved in the work on the atomic bomb, then in the work on the hydrogen bomb, but most of my life has been spent in more theoretical realms. My friend Otto Frisch, the discoverer of the possibility of chain reactions from fission, in an article in *The Bulletin of the Atomic Scientists* describing his first impressions of Los Alamos upon arriving there from embattled Britain, wrote:

"Certainly I have never found such a concentration of interesting people in one place. In the evening I felt I could walk into any house at random and would find congenial people engaged in music making or in stimulating debate. . . . I also met Stan Ulam early on, a brilliant Polish topologist with a charming French wife. At once he told me that he was a pure mathematician who had sunk so low that his latest paper actually contained numbers with decimal points!"

Little has been written about the lives of the people responsible for so much in science and in the birth of the nuclear age and the space age: von Neumann, Fermi, and numerous other mathematicians and physicists. But here I want to recount also the more abstract and philosophically decisive influences which came from mathematics itself. Names like Stefan Banach, G. D. Birkhoff, and David Hilbert are virtually unknown to the general public, and yet it is these men, along with Einstein, Fermi and a few others equally famous, who were indispensable to what twentieth-century science has accomplished.

PART I

Becoming
a Mathematician
in Poland

CHAPTER 1

Childhood
1909–1927

MY father, Jozef Ulam, was a lawyer. He was born in
Lwów, Poland, in 1877. At the time of his birth the
city was the capital of the province of Galicia, part
of the Austro-Hungarian Empire. When I was born in 1909
this was still true.

His father, my grandfather, was an architect and a building contractor. I understand that my great-grandfather had
come to Lwów from Venice.

My mother, Anna Auerbach, was born in Stryj, a small
town some sixty miles south of Lwów, near the Carpathian
Mountains. Her father was an industrialist who dealt in steel
and represented factories in Galicia and Hungary.

One of my earliest memories is of sitting on a window
sill with my father and looking out at a street on which there
was a great parade honoring the Crown Prince, who was
visiting Lwów. I was not quite three years old.

I remember when my sister was born. I was told a little

girl had arrived, and I felt—it is hard to describe—somehow grown up. I was three.

When I was four, I remember jumping around on an oriental rug looking down at its intricate patterns. I remember my father's towering figure standing beside me, and I noticed that he smiled. I felt, "He smiles because he thinks I am childish, but I know these are curious patterns." I did not think in those very words, but I am pretty certain that it was not a thought that came to me later. I definitely felt, "I know something my father does not know. Perhaps I know better than my father."

I also have the memory of a trip to Venice with the family. We were on a vaporetto on a canal, and I had a balloon which fell overboard. As it bobbed along the side of the boat, my father tried to fish it out with the crooked end of his walking stick but failed. I was consoled by being allowed to select a souvenir model of a gondola made of Venetian beads and still remember the feeling of pride at being given such a task.

I remember the beginning of the first World War. As a boy, I was a Central Powers patriot when Austria, Germany, and Bulgaria—the "Central Powers"—were fighting against France, England, Russia, and Italy. Most of the Polish-speaking people were nationalistic and anti-Austrian, but nevertheless, at about the age of eight I wrote a little poem about the great victories of the Austrian and German armies.

Early in 1914, the Russian troops advanced into Galicia and occupied Lwów. My family left, taking refuge in Vienna. There I learned German, but my native language—the language we spoke at home—was Polish.

We lived in a hotel across from St. Stephen's cathedral. The strange thing is that even though I visited Vienna many times afterwards, I did not actually recognize this building again until one day in 1966 while I was walking through the streets with my wife. Perhaps because we were talking about my childhood I suddenly remembered it and pointed it out to her. With this a number of other memories buried for over fifty years surfaced.

On the same visit, while walking through the Prater gardens, the sight of an outdoor café suddenly brought back the memory of how I had once choked in the wind with a sort of asthmatic reaction in front of that very café—a feeling that I was not to experience again until many years later in Madison, Wisconsin. Curiously the subsequent sensation did not make me recall the childhood episode. It is only when I was at that very spot many years later that this sensory memory returned as a result of the visual association.

I will not try to describe the mood of Vienna as seen through the eyes of a six-year-old. I wore a sort of military cap; when an officer saluted me on Kärntner Strasse (one of the main streets of Vienna) I remember vividly that I was absolutely delighted. But when somebody mentioned that the United States would have ten thousand airplanes (there was such a rumor) I began to have doubts about the victory of the Central Powers.

At about this time in Vienna I learned to read. Like so much of learning throughout my life, at first it was an unpleasant—a difficult, somewhat painful experience. After a while, everything fell into place and became easy. I remember walking the streets reading all the signs aloud with great pleasure, probably annoying my parents.

My father was an officer in the Austrian Army attached to military headquarters, and we traveled frequently. For a while we lived in Märisch Ostrau, and I went to school there for a time. In school we had to learn the multiplication tables, and I found learning arithmetic mildly painful. Once I was kept home with a cold just as we were at six times seven. I was sure that the rest of the class would be at twelve times fifteen by the time I went back. I think I went to ten times ten by myself. The rest of the time I had tutors, for we traveled so much it was not possible to attend school regularly.

I also remember how my father would sometime read to me from a children's edition of Cervantes' *Don Quixote*. Episodes that now seem only mildly funny to me, I considered hilarious. I thought the description of Don Quixote's fight with the windmills the funniest thing imaginable.

These are visual pictures, not nostalgic really but bearing a definite taste, and they leave a definite flavor of associations in the memory. They carry with them a consciousness of different intensities, different colors, different compositions, mixed with feelings which are not explicit—of well-being or of doubt. They certainly play simultaneously on many physically separate parts in the brain and produce a feeling perhaps akin to a melody. It is a reconstruction of how I felt. People often retain these random pictures, and the strange thing is that they persist throughout one's life.

Certain scenes are easier of access, but there are probably many other impressions which continue to exist: Experiments have re-created certain scenes from the past when areas of a patient's brain were touched with a needle during an operation. The scenes that can be summoned up from one's memory at will have a color or flavor which does not seem to change with time. Their re-creation by recollection does not seem to change them or refresh them. As far as I can tell when I try to observe in myself the chain of syllogisms initiated by these impressions, they are quite analogous now as to what they were when I was little. If I look now at an object, like a chair, or a tree, or a telegraph wire, it initiates a train of thought. And it seems to me that the succession of linked memories are quite the same as those I remember when I was five or six. When I look at a telegraph wire, I remember very well it gave me a sort of abstract or mathematical impulse. I wondered what else could do that. It was an attempt at generalization.

Perhaps the store of memory in the human brain is to a large extent already formed at a very early age, and external stimuli initiate a process of recording and classifying the impressions along channels which exist in large numbers in very early childhood.

To learn how things are filed in the memory, it obviously helps to analyze one's thoughts. To understand how one understands a text, or a new method, or a mathematical proof, it is interesting to try to consciously perceive the temporal order and the inner logic. Professionals or even interested amateurs have not done enough in this area to judge by what I have read on the nature of memory. It seems to me that more could be done to elicit even in part the nature of associations, with computers providing the means for experimentation. Such a study would have to involve a gradation of notions, of symbols, of classes of symbols, of classes of classes, and so on, in the same way that the complexity of mathematical or physical structures is investigated.

There must be a trick to the train of thought, a recursive formula. A group of neurons starts working automatically, sometimes without external impulse. It is a kind of iterative process with a growing pattern. It wanders about in the brain, and the way it happens must depend on the memory of similar patterns.

Very little is known about this. Perhaps before a hundred years have passed this will all be part of a fascinating new science. It was not so long ago that scientists like John von Neumann began to examine analogies between the operation of the brain and that of the computer. Earlier, people had thought the heart was the seat of thought; then the role of the brain became more evident. Perhaps it actually depends on all the senses.

We are accustomed to think of thinking as a linear experience, as when we say "train" of thought. But subconscious thinking may be much more complicated. Just as one has simultaneous visual impressions on the retina, might there not be simultaneous, parallel, independently originated, abstract impressions in the brain itself? Something goes on in our heads in processes which are not simply strung out on one line. In the future, there might be a theory of a memory search, not by one sensor going around, but perhaps more like several searchers looking for someone lost in a forest. It is a problem of pursuit and of search—one of the greatest areas of combinatorics.

What happens when one suddenly remembers a forgotten word or name? What does one do when one tries to remember it? Subconsciously something is turning. More than one route is followed: one tries by sound or letters, long words or short words. That must mean that the word is filed in multiple storage. If it were only in one place there would be no way to recover it. Time is a parameter, too, and although in the conscious there seems to be only one time, there may be many in the subconscious. Then there is the mechanism of synthesizer or summarizer. Could one introduce an automatic search system, an ingenious system which does not go through everything but scans the relevant elements?

But I have digressed enough in these observations on memory. Let me now return to this account of my life. I only wish that I could have some of Vladimir Nabokov's ability to evoke panoramas of memories from a few pictures of the past. Indeed one can say that an artist depicts the essential functions or properties of a whole set of impressions on the retina. It is these that the brain summarizes and stores in the memory, just as a caricaturist can convey the essentials of a face with just a few strokes. Mathematically speaking, these are the global characteristics of the function or the figure of a set of points. In this more prosaic account I will describe merely the more formal points.

In 1918 we returned to Lwów, which had become part of the newly formed Republic of Poland. In November of that year the Ukrainians besieged the city, which was defended by a small number of Polish soldiers and armed civilians. Our house was in a relatively safe part of town, even though occasional artillery shells struck nearby. Because our house was safer, many of our relatives came to stay with us. There must have been some thirty of them, half being children. There were not nearly enough beds, of course, and I remember people sleeping everywhere on rolled rugs on the floor. During the shelling we had to go to the basement. I still remember insisting on tying my shoes while my mother was pressing me to hurry downstairs. For the adults it must have been a strenuous time to say the least, but not for us. Strangely enough, my memories of these days are of the fun I had playing, hiding, learning card games with the children for the two weeks before the siege was lifted with the arrival of another Polish army from France. This broke the ring of besiegers. For children wartime memories are not always traumatic.

During the Polish-Russian war in 1920 the city was threatened again. Budenny's cavalry penetrated to within fifty miles, but Pilsudski's victory on the Warsaw front saved the southern front and the war ended.

At the age of ten in 1919 I passed the entrance examination to the gymnasium. This was a secondary school patterned after the German gymnasia and the French lycées. Instruction usually took eight years. I was an A student, except in penmanship and drawing, but did not study much.

One of the gaps in my education was in chemistry. We did not have much of it in school and fifty years later, now that I am interested in biology, this handicaps me in my studies of elementary biochemistry.

About this time I also discovered that I did not have quite normal binocular vision. It happened in the following way: the boys in the class had been lined up for an eye examination. Awaiting my turn to read the charts, I covered my eyes with my hand. I noticed with horror that I could only read the largest letters with my right eye. This made me afraid that I would be kept after school, so I memorized the letters. I think it was the first time in my life when I consciously cheated. When my turn came I "read" satisfactorily and was let off, but I knew my eyes were different, one was myopic. The other, normal, later became presbyotic. This

condition, rather rare but well known, is apparently hereditary. I still have never worn glasses, although I have to bend
close to the printed text to read with my myopic eye. I am
not normally aware which eye I use; once later in life a doctor in Madison told me that this condition is sometimes better than normal, for one or the other eye is resting while the
other is in use. I wonder if my peculiar eyesight, in addition
to affecting my reading habits, may also have affected my
habits of thought.

When I try to remember how I started to develop my interest in science I have to go back to certain pictures in a
popular book on astronomy I had. It was a textbook called
Astronomy of Fixed Stars, by Martin Ernst, a professor of astronomy at the University of Lwów. In it was a reproduction
of a portrait of Sir Isaac Newton. I was nine or ten at the
time, and at that age a child does not react consciously to the
beauty of a face. Yet I remember distinctly that I considered
this portrait—especially the eyes—as something marvelous.
A mixture of physical attraction and a feeling of the mysterious emanated from his face. Later I learned it was the Geoffrey Kneller portrait of Newton as a young man, with hair to
his shoulders and an open shirt. Other illustrations I distinctly remember were of the rings of Saturn and of the belts
of Jupiter. These gave me a certain feeling of wonder, the
flavor of which is hard to describe since it is sometimes associated with nonvisual impressions such as the feeling one
gets from an exquisite example of scientific reasoning. But it
reappears, from time to time, even in older age, just as a familiar scent will reappear. Occasionally an odor will come
back, bringing coincident memories of childhood or youth.

Reading descriptions of astronomical phenomena today
brings back to me these visual memories, and they reappear
with a nostalgic (not melancholy but rather pleasant) feeling, when new thoughts come about or a new desire for
mental work suddenly emerges.

The high point of my interest in astronomy and an un-
forgettable emotional experience came when my uncle Szy-

mon Ulam gave me a little telescope. It was one of the copper- or bronze-tube variety and, I believe, a refractor with a two-inch objective.

To this day, whenever I see an instrument of this kind in antique shops, nostalgia overcomes me, and after all these decades my thoughts still turn to visions of the celestial wonders and new astronomical problems.

At that time, I was intrigued by things which were not well understood—for example, the question of the shortening of the period of Encke's comet. It was known that this comet irregularly and mysteriously shortens its three-year period of motion around the sun. Nineteenth-century astronomers made several attempts to account for this as being caused by friction or by the presence of some new invisible body in space. It excited me that nobody really knew the answer. I speculated whether the $1/r^2$ law of attraction of Newton was not quite exact. I tried to imagine how it could affect the period of the comet if the exponent was slightly different from 2, imagining what the result would be at various distances. It was an attempt to calculate, not by numbers and symbols, but by almost tactile feelings combined with reasoning, a very curious mental effort.

No star could be large enough for me. Betelgeuse and Antares were believed to be much larger than the sun (even though at the time no precise data were available) and their distances were given, as were parallaxes of many stars. I had memorized the names of constellations and the individual Arabic names of stars and their distances and luminosities. I also knew the double stars.

In addition to the exciting Ernst book another, entitled *Planets and the Conditions of Life on Them,* was strange. Soon I had some eight or ten astronomy books in my library, including the marvelous Newcomb-Engelmann *Astronomie* in German. The Bode-Titius formula or "law" of planetary distances also fascinated me, inspiring me to become an astronomer or physicist. This was about the time when, at the age of eleven or so, I inscribed my name in a notebook, "S.

Ulam, astronomer, physicist and mathematician." My love for astronomy has never ceased; I believe it is one of the avenues that brought me to mathematics.

From today's perspective Lwów may seem to have been a provincial city, but this is not so. Frequent lectures by scientists were held for the general public, in which such topics as new discoveries in astronomy, the new physics and the theory of relativity were covered. These appealed to lawyers, doctors, businessmen, and other laymen.

Other popular lecture topics were Freud and psychoanalysis. Relativity theory was, of course, much more difficult.

Around 1919–1920 so much was written in newspapers and magazines about the theory of relativity that I decided to find out what it was all about. I went to some of the popular talks on relativity. I did not really understand any of the details, but I had a good idea of the main thrust of the theory. Almost like learning a language in childhood, one develops the ability to speak it without knowing anything about grammar. Curiously enough, it is possible even in the exact sciences to have an idea of the gist of something without having a complete understanding of the basics. I understood the schema of special relativity and even some of its consequences without being able to verify the details mathematically. I believe that so-called understanding is not a yes-or-no proposition. But we don't yet have the technique of defining these levels or the depth of the knowledge of reasons.

This interest became known among friends of my father, who remarked that I "understood" the theory of relativity. My father would say, "The little boy seems to understand Einstein!" This gave me a reputation I felt I had to maintain, even though I knew that I did not genuinely understand any of the details. Nevertheless, this was the beginning of my reputation as a "bright child." This encouraged me to further study of popular science books—an experience

I am sure is common to many children who later grow up to be scientists.

How a child acquires the habits and interests which play such a decisive role in determining his future has not been sufficiently investigated. "Plagiarism"—the mysterious ability of a child to imitate or copy external impressions such as the mother's smile—is one possible explanation. Another is inborn curiosity: why does one seek new experiences instead of merely reacting to stimuli?

Inclinations may be part of the inherited system of connections in the brain, a genetic trait that may not even depend on the physical arrangement of neurons. Apparently headaches are related to the ease with which blood circulates in the brain, which depends on whether the blood vessels are wide or narrow. Perhaps it is the "plumbing" that is important, rather than the arrangement of the neurons normally associated with the seat of thinking.

Another determining factor may be initial accidents of success or failure in a new pursuit. I believe that the quality of memory develops similarly as a result of initial accidents, random external influences, or a lucky combination of the two.

Consider the talent for chess, for example. José Capablanca learned the game at the age of six by watching his father and uncle play. He developed the ability to play naturally, effortlessly, the way a child learns to speak as compared with the struggles adults have in learning new subjects. Other famous chess players also first became interested by watching their relatives play. When they tried, perhaps a chance initial success encouraged them to pursue. Nothing succeeds like success, it is well known, especially in early youth.

I learned chess from my father. He had a little paperbound book on the subject and used to tell me about some of the famous games it described. The moves of the knight fascinated me, especially the way two enemy pieces can be threatened simultaneously with one knight. Although it is a simple stratagem, I thought it was marvelous, and I have loved the game ever since.

Could the same process apply to the talent for mathematics? A child by chance has some satisfying experiences with numbers; then he experiments further and enlarges his memory by building up his store of experiences.

I had mathematical curiosity very early. My father had in his library a wonderful series of German paperback books—*Reklam,* they were called. One was Euler's *Algebra.* I looked at it when I was perhaps ten or eleven, and it gave me a mysterious feeling. The symbols looked like magic signs; I wondered whether one day I could understand them. This probably contributed to the development of my mathematical curiosity. I discovered by myself how to solve quadratic equations. I remember that I did this by an incredible concentration and almost painful and not-quite-conscious effort. What I did amounted to completing the square in my head without paper or pencil.

In high school, I was stimulated by the notion of the problem of the existence of odd perfect numbers. An integer is perfect if it is equal to the sum of all its divisors including one but not itself. For instance: $6 = 1 + 2 + 3$ is perfect. So is $28 = 1 + 2 + 4 + 7 + 14$. You may ask: does there exist a perfect number that is odd? The answer is unknown to this day.

In general, the mathematics classes did not satisfy me. They were dry, and I did not like to have to memorize certain formal procedures. I preferred reading on my own.

At about fifteen I came upon a treatise on the infinitesimal calculus in a book by Gerhardt Kowalevski. I did not have enough preparation in analytic geometry or even in trigonometry, but the idea of limits, the definitions of real numbers, the notion of derivatives and integration puzzled and excited me greatly. I decided to read a page or two a day and attempt to learn the necessary facts about trigonometry and analytic geometry from other books.

I found two other books in a secondhand bookstore.

These intrigued and fascinated me more than anything else for many years to come: Sierpinski's *Theory of Sets* and a monograph on number theory. At the age of seventeen I knew as much or more elementary number theory than I do now.

I also read a book by the mathematician Hugo Steinhaus entitled *What Is and What Is Not Mathematics* and in Polish translation Poincaré's wonderful *La Science et l'Hypothèse, La Science et la Méthode, La Valeur de la Science,* and his *Dernières Pensées.* Their literary quality, not to mention the science, was admirable. Poincaré molded portions of my scientific thinking. Reading one of his books today demonstrates how many wonderful truths have remained, although everything in mathematics has changed almost beyond recognition and in physics perhaps even more so. I admired Steinhaus's book almost as much, for it gave many examples of actual mathematical problems.

The mathematics taught in school was limited to algebra, trigonometry, and the very beginning of analytic geometry. In the seventh and eighth classes, where the students were sixteen and seventeen, there was a course on elementary logic and a survey of history of philosophy. The teacher, Professor Zawirski, was a real scholar, a lecturer at the University and a very stimulating man. He gave us glimpses of recent developments in advanced modern logic. Having studied Sierpinski's books on the side, I was able to engage him in discussions of set theory during recess and in his office. I was working on some problems on transfinite numbers and on the problem of the continuum hypothesis.

I also engaged in wild mathematical discussions, formulating vast and new projects, new problems, theories and methods bordering on the fantastic, with a boy named Metzger, some three or four years my senior. He had been directed toward me by friends of my father who knew that he too had a great interest in mathematics. Metzger was short, rotund, blondish, a typical liberated ghetto Jew. Later I saw a youthful portrait of Heine which reminded me of his face. People of his type can still be found occasionally. They exhibit amateurism, even about the very foundations of arithmetic. We discussed "an iterative calculus" on the basis of practically no knowledge of the existing mathematical material. He was "crazy" and full of the urge to innovate which is so Jewish. Stefan Banach once pointed out that it is characteristic of certain Jews always to try to change the established scheme of things—Jesus, Marx, Freud, Cantor. On a very small scale Metzger showed this tendency. Had he had a better education he might have done good things. He obviously came from a very poor family and his Polish had a strong, guttural accent. After a few months he abruptly vanished from my ken. This is the first time I have thought about him in all these years. Perhaps he is alive. This memory of Metzger and our discussions brings back the very smell and color of the "abstractions" we exchanged.

Strangely enough, at this youthful and immature age I was also occasionally trying to analyze my own thinking processes. I tried to make myself more aware of them by periodically going back every few seconds to see what it was that molded the train of thought. Needless to say, I was fully aware of the fact that there is a danger in indulging too much and too frequently in such introspection.

So far, the image I had formed of astronomers and scientists, and of mathematicians in particular, came almost exclusively from my reading. I got my first "live" impressions when I went to a series of popular mathematics lectures in 1926. On successive days there were talks by Hugo Steinhaus, Stanislaw Ruziewicz, Stefan Banach, and perhaps others. My first surprise was to discover how young they were. Having heard and read of their achievements I really expected bearded old scholars. I listened avidly to their talks. Young as I was, my impression of Banach was that here was a homespun genius. This first impression—deepened, enriched, and transformed, of course—remained during my subsequent long acquaintance, collaboration, and friendship with him.

CHILDHOOD

Then in 1927, Zawirski told me a congress of mathematicians was to take place in Lwów and foreign scholars had been invited. He added that a youthful and extremely brilliant mathematician named John von Neumann was to give a lecture. This was the first time I heard the name. Unfortunately, I could not attend these lectures for I was in the midst of my own matriculation examinations at the Gymnasium.

Still, my interests in science did not take all of my time. I avidly read Polish literature, as well as writers as diverse as Tolstoy, Jules Verne, Karl May, H. G. Wells, and Anatole France. As a boy I preferred biographies and adventure stories.

Besides these more cerebral activities, I engaged actively in sports. Beginning at about fourteen I played various positions in soccer with my classmates: goalie, right forward and others. I started playing tennis, too, and was active in track and field.

After school I played cards with my classmates. We played bridge and a simple variety of poker for small stakes. In poker the older boys won most of the time. One of the abilities that apparently does not decrease but rather improves with age is a primitive type of elementary shrewdness. I played chess also, two or three times a week. Although I don't think I ever had too much talent for the game, I certainly had a more than average feeling for positions, and I probably was one of the best players in my group. Like mathematics, chess is one of the things where constant practice, constant thinking, and imagining, and studying are necessary to achieve a mastery of the game.

In 1927 I passed my three-day matriculation examinations and a period of indecision began. The choice of a future career was not easy. My father, who had wanted me to become a lawyer so I could take over his large practice, now recognized that my inclinations lay in other directions. Besides, there was no shortage of lawyers in Lwów. The thought of a university career was attractive, but professorial positions were rare and hard to obtain, especially for people with Jewish backgrounds like myself. Consequently, I looked for a course of studies which would lead to something practical and at the same time would be connected with science. My parents urged me to become an engineer, and so I applied for admission at the Lwów Polytechnic Institute as a student of either mechanical or electrical engineering.

CHAPTER 2

Student Years

1927–1933

IN the fall of 1927 I began attending lectures at the Polytechnic Institute in the Department of General Studies, because the quota of Electrical Engineering already was full. The level of the instruction was obviously higher than that at high school, but having read Poincaré and some special mathematical treatises, I naively expected every lecture to be a masterpiece of style and exposition. Of course, I was disappointed.

As I knew many of the subjects in mathematics from my studies, I began to attend a second-year course as an auditor. It was in set theory and given by a young professor fresh from Warsaw, Kazimir Kuratowski, a student of Sierpinski, Mazurkiewicz, and Janiszewski. He was a freshman professor, so to speak, and I a freshman student. From the very first lecture I was enchanted by the clarity, logic, and polish of his exposition and the material he presented. From the beginning I participated more actively than most of the older students in discussions with Kuratowski, since I knew something of the subject from having read Sierpinski's book. I think he quickly noticed that I was one of the better students; after class he would give me individual attention. This is how I started on my career as a mathematician, stimulated by Kuratowski.

Soon I could answer some of the more difficult questions in the set theory course, and I began to pose other problems. Right from the start I appreciated Kuratowski's patience and generosity in spending so much time with a novice. Several times a week I would accompany him to his apartment at lunch time, a walk of about twenty minutes, during which I asked innumerable mathematical questions. Years later, Kuratowski told me that the questions were sometimes significant, often original, and interesting to him.

My courses included mathematical analysis, calculus, classical mechanics, descriptive geometry, and physics. Between classes, I would sit in the offices of some of the mathematics instructors. At that time I was perhaps more eager than at any other time in my life to do mathematics to the exclusion of almost any other activity.

It was there that I first met Stanislaw Mazur, who was a young assistant at the University. He came to the Polytechnic Institute to work with Orlicz, Nikliborc and Kaczmarz, who were a few years his senior.

In conversations with Mazur I began to learn about problems in analysis. I remember long hours of sitting at a desk and thinking about the questions which he broached to me and discussed with the other mathematicians. Mazur introduced me to advanced ideas of real variable function theory and the new functional analysis. We discussed some of the more recent problems of Banach, who had developed a new approach to this theory.

Banach himself would appear occasionally, even though his main work was at the University. I met him during this first year, but our acquaintance began in a more meaningful, intimate, and intellectual sense a year or two later.

Several other mathematicians could frequently be seen in these offices. Stozek, cheerful, rotund, short, and completely bald, was Chairman of the Department of General Studies. The word *stozek* means "a cone" in Polish; he looked more like a sphere. Always in good humor and joking incessantly, he loved to consume frankfurters liberally smeared with horseradish, a dish which he maintained cured melancholy. (Stozek was one of the professors murdered by the Germans in 1941.)

Antoni Lomnicki, a mathematician of aristocratic features who specialized in probability theory and its applications to cartography, had office hours in these rooms. (He too was murdered by the Germans in Lwów in 1941.) His nephew, Zbigniew Lomnicki, later became my good friend and mathematical collaborator.

Kaczmarz, tall and thin (who later was killed in military service in 1940), and Nikliborc, short and rotund, managed the exercise sections of the large calculus and differential equations courses. They were often seen together and reminded me of Pat and Patachon, two contemporary comic film actors.

I did not feel I was a regular student in the sense that one may have to study subjects one is not especially interested in. On the other hand, after all these years, I still do not feel much like an accomplished professional mathematician. I like to try new approaches and, being an optimist by nature, hope they will succeed. It has never occurred to me to question whether a mental effort will be wasted or whether to "husband" my mental capital.

At the beginning of the second semester of my freshman year, Kuratowski told me about a problem in set theory that involved transformations of sets. It was connected with a well-known theorem of Bernstein: if $2A = 2B$, then $A = B$, in the arithmetic sense of infinite cardinals. This was the first problem on which I really spent arduous hours of thinking. I thought about it in a way which now seems mysterious to me, not consciously or explicitly knowing what I was aiming at. So immersed in some aspects was I, that I did not have a conscious overall view. Nevertheless, I managed to show by means of a construction how to solve the problem, devising a method of representing by graphs the decomposition of sets and the corresponding transformations. Unbelievably, at the time I thought I had invented the very idea of graphs.

I wrote my first paper on this in English, which I knew better than German or French. Kuratowski checked it and the short paper appeared in 1928 in *Fundamenta Mathematicae,* the leading Polish mathematical journal which he edited. This gave me self-confidence.

I still was not certain what career or course of work I should pursue. The practical chances of becoming a professor of mathematics in Poland were almost nil—there were few vacancies at the University. My family wanted me to learn a profession, and so I intended to transfer to the Department of Electrical Engineering for my second year. In this field the chance of making a living seemed much better.

Before the end of the year Kuratowski mentioned in a lecture another problem in set theory. It was on the existence of set functions which are "subtractive" but not completely countably additive. I remember pondering the question for weeks. I can still feel the strain of thinking and the number of attempts I had to make. I gave myself an ultimatum. If I could solve this problem, I would continue as a mathematician. If not, I would change to electrical engineering.

After a few weeks I found a way to achieve a solution. I ran excitedly to Kuratowski and told him about my solution, which involved transfinite induction. Transfinite induction had been used by mathematical workers many times in other connections; however, I believe that the way in which I used it was novel.

I think Kuratowski took pleasure in my success, encouraging me to continue in mathematics. Before the end of my first college year I had written my second paper, which

Kuratowski presented to *Fundamenta*. Now, the die was cast. I began to concentrate on the "impractical" possibilities of an academic career. Most of what people call decision making occurs for definite reasons. However, I feel that for most of us what is ultimately called a "decision" is a sort of vote taken in the subconscious, in which the majority of the reasons favoring the decision win out.

During the summer of 1928 when I took a trip to the Baltic coast of Poland, Kuratowski invited me to visit him on the way at his summer place near Warsaw. It was an elegant villa with a tennis court. Kuratowski was quite good at tennis in those days, and this surprised me since his figure was anything but athletic.

On the six-hour train ride from Lwów to Warsaw I thought almost without interruption about problems in set theory with the idea of presenting something that would interest him. I was thinking of ways to disprove the continuum hypothesis, a famous unsolved problem in foundations of set theory and mathematics formulated by Georg Cantor, the creator of set theory. My presentation was vague, and Kuratowski soon detected this. Nevertheless, we discussed its ramifications, and so I went on to Zoppot with my self-confidence intact.

Alfred Tarski, now a celebrated logician and professor at Berkeley, was a friend of Kuratowski from Warsaw, who occasionally visited Lwów. He was already known internationally as a logician, but his work in the foundations of mathematical logic and set theory was also important. He had been a candidate for a chair of philosophy that was vacant at the University of Lwów. The chair went instead to another logician, Leon Chwistek, an accomplished painter and author of philosophical treatises, a brother-in-law of Steinhaus, and well known for many eccentricities. (He died in Moscow during the war.) Years later in Cambridge, I happened to mention Chwistek to Alfred North Whitehead. In the course of the conversation I said, "Very strange, he was a painter too!" Whereupon Whitehead laughed out loud, clapped his hands and exclaimed: "How British of you to say that being a painter is strange." Mrs. Whitehead joined in the laughter. A very good biography of Chwistek by Estriecher has recently appeared in Poland. It is a fascinating account of the intellectual and artistic life of Cracow and Lwów from 1910 to 1946.

One of my early contacts with Tarski was a result of my second paper. In it I had proved a theorem on ideals of sets in set theory. (Marshall Stone later proved another version of this same theorem.) My note in *Fundamenta* also showed the possibility of defining a finitely additive measure with two values, 0 to 1, and established a maximum prime ideal for subsets in the infinite set. In a very long paper which appeared a year later, Tarski got the same result. After Kuratowski pointed out to him that it followed from my theorem, Tarski acknowledged this in a footnote. In view of my youth, this seemed to me a little victory—an acknowledgment of my mathematical presence.

There was a feeling among some mathematicians that logic is not "real" mathematics, but merely a preparatory and somewhat alien art. Today, this feeling is disappearing as a result of many concrete mathematical advances made by the methods of formal logic.

During the second year of studies I decided to audit a course in theoretical physics given by Professor Wojciech Rubinowicz, a leading Polish theoretician and a former student and collaborator of the famous Munich physicist Sommerfeld.

I attended his masterly lectures on electromagnetism and took part in a seminar he led on group theory and quantum theory for advanced students. We used Hermann Weyl's *Gruppen Theorie und Quantum Mechanik*. It was impressive to see the high level of mathematics involved in the study of Maxwell's equations and in the theory of electricity which made up its first part. Even though much of it was above my head technically, I managed to do a lot of reading on the side. I read popular accounts of theoretical physics in statistical mechanics, in the theory of gases and the theory of relativity, and on electricity and magnetism.

During the winter, Rubinowicz fell ill and asked me (although I was the youngest member of the class) to conduct a few sessions during his absence. I remember to this day how I struggled with the unfamiliar and difficult material of Weyl's book. This was my first active participation in the area of physics.

The mathematics offices of the Polytechnic Institute continued to be my hangout. I spent mornings there, every day of the week, including Saturdays. (Saturdays were not considered to be part of the weekend then; classes were held on Saturday mornings.)

Mazur appeared often, and we started our active collaboration on problems of function spaces. We found a solution to a problem involving infinitely dimensional vector spaces. The theorem we proved—that a transformation preserving distances is linear—is now part of the standard treatment of the geometry of function spaces. We wrote a paper which was published in the *Compte-Rendus* of the French Academy.

It was Mazur (along with Kuratowski and Banach) who introduced me to certain large phases of mathematical thinking and approaches. From him I learned much about the attitudes and psychology of research. Sometimes we would sit for hours in a coffee house. He would write just one symbol or a line like $y = f(x)$ on a piece of paper, or on the marble table top. We would both stare at it as various thoughts were suggested and discussed. These symbols in front of us were like a crystal ball to help us focus our concentration. Years later in America, my friend Everett and I often had similar sessions, but instead of a coffee house they were held in an office with a blackboard.

Mazur's forte was making what he called "observations and remarks." These stated—usually in a concise and precise form—some properties of notions. Once made, they were perhaps not so difficult to verify, for sometimes they were peripheral to the usual formulations and had gone unnoticed. They were often decisive in solving problems.

In a conversation in the coffee house, Mazur proposed the first examples of infinite mathematical games. I remember also (it must have been sometime in 1929 or 1930) that he raised the question of the existence of automata which would be able to replicate themselves, given a supply of some inert material. We discussed this very abstractly, and some of the thoughts which we never recorded were actually precursors of theories like that of von Neumann on abstract automata. We speculated frequently about the possibility of building computers which could perform exploratory numerical operations and even formal algebraical work.

I have mentioned that I first saw Banach at a series of mathematics lectures when I was in high school. He was then in his middle thirties, but contrary to the impression given to very young people by men fifteen or twenty years their senior, to me he appeared to be very youthful. He was tall, blond, blue-eyed, and rather heavy-set. His manner of speaking struck me as direct, forceful, and perhaps too simple-minded (a trait which I later observed was to some extent consciously forced). His facial expression was usually one of good humor mixed with a certain skepticism.

Banach came from a poor family, and he had very little conventional schooling at first. He was largely self-taught when he arrived at the Polytechnical Institute. It is said that Steinhaus accidentally discovered his talent when he overheard a mathematical conversation between two young students sitting on a park bench. One was Banach, the other Nikodym, now recently retired as professor of mathematics at Kenyon College. Banach and Steinhaus were to become the closest of collaborators and the founders of the Lwów school of mathematics.

Banach's knowledge of mathematics was broad. His contributions were in the theory of functions of real variables, set theory and, above all, functional analysis, the theory of spaces of infinitely many dimensions (the points of these spaces being functions or infinite series of numbers). They include some of the most elegant results. He once told me that as a young man he knew the three volumes of Darboux's *Differential Geometry*.

I attended only a few of Banach's lectures. I especially remember some on the calculus of variations. In general, his lectures were not too well prepared; he would occasionally make mistakes or omissions. It was most stimulating to watch him work at the blackboard as he struggled and invariably managed to pull through. I have always found such a lecture more stimulating than the entirely polished ones where my attention would lapse completely and would revive only when I sensed that the lecturer was in difficulty.

Beginning with the third year of studies, most of my mathematical work was really started in conversations with Mazur and Banach. And according to Banach some of my own contributions were characterized by a certain "strangeness" in the formulation of problems and in the outline of possible proofs. As he told me once some years later, he was surprised how often these "strange" approaches really worked. Such a statement, coming from the great master to a young man of twenty-eight, was perhaps the greatest compliment I have received.

In mathematical discussions, or in short remarks he made on general subjects, one could feel almost at once the great power of his mind. He worked in periods of great intensity separated by stretches of apparent inactivity. During the latter his mind kept working on selecting the statements, the sort of alchemist's probe stones that would best serve as focal theorems in the next field of study.

He enjoyed long mathematical discussions with friends and students. I recall a session with Mazur and Banach at the Scottish Café which lasted seventeen hours without interruption except for meals. What impressed me most was the way he could discuss mathematics, reason about mathematics, and find proofs in these conversations.

Since many of these discussions took place in neighborhood coffee houses or little inns, some mathematicians also dined there frequently. It seems to me now the food must have been mediocre, but the drinks were plentiful. The tables had white marble tops on which one could write with a pencil, and, more important, from which notes could be easily erased.

There would be brief spurts of conversation, a few lines would be written on the table, occasional laughter would come from some of the participants, followed by long periods of silence during which we just drank coffee and stared vacantly at each other. The café clients at neighboring tables must have been puzzled by these strange doings. It is such persistence and habit of concentration which somehow becomes the most important prerequisite for doing genuinely creative mathematical work.

Thinking very hard about the same problem for several hours can produce a severe fatigue, close to a breakdown. I never really experienced a breakdown, but have felt "strange inside" two or three times during my life. Once I was thinking hard about some mathematical constructions, one after the other, and at the same time trying to keep them all simultaneously in my mind in a very conscious effort. The concentration and mental effort put an added strain on my nerves. Suddenly things started going round and round, and I had to stop.

These long sessions in the cafés with Banach, or more often with Banach and Mazur, were probably unique. Collaboration was on a scale and with an intensity I have never seen surpassed, equaled or approximated anywhere—except perhaps at Los Alamos during the war years.

Banach confided to me once that ever since his youth he had been especially interested in finding proofs—that is, demonstrations of conjectures. He had a subconscious system for finding hidden paths—the hallmark of his special genius.

After a year or two Banach transferred our daily sessions

from the Café Roma to the "Szkocka" (Scottish Café) just across the street. Stozek was there every day for a couple of hours, playing chess with Nikliborc and drinking coffee. Other mathematicians surrounded them and kibitzed.

Kuratowski and Steinhaus appeared occasionally. They usually frequented a more genteel teashop that boasted the best pastry in Poland.

It was difficult to outlast or outdrink Banach during these sessions. We discussed problems proposed right there, often with no solution evident even after several hours of thinking. The next day Banach was likely to appear with several small sheets of paper containing outlines of proofs he had completed in the meantime. If they were not polished or even not quite correct, Mazur would frequently put them in a more satisfactory form.

Needless to say such mathematical discussions were interspersed with a great deal of talk about science in general (especially physics and astronomy), university gossip, politics, the state of affairs in Poland; or, to use one of John von Neumann's favorite expressions, the "rest of the universe." The shadow of coming events, of Hitler's rise in Germany, and the premonition of a world war loomed ominously.

Banach's humor was ironical and sometimes tinged with pessimism. For a time he was dean of the Faculty of Science and had to attend various committee meetings. He tried to avoid all such activities, as much as he could, and once he told me, "*Wiem gdzie nie będę* [I know where I won't be]," his way of saying that he did not intend to attend a dull meeting.

Banach's faculty for proposing problems illuminating whole sections of mathematical disciplines was very great, and his publications reflect only a part of his mathematical powers. The diversity of his mathematical interests surpassed that shown in his published work. His personal influence on other mathematicians in Lwów and in Poland was very strong. He stands out as one of the main figures of this remarkable period between the wars when so much mathematical work was accomplished.

I have had no precise knowledge of his life and work from the outbreak of the war to his premature death in the fall of 1945. From fragments of information obtained later, we learned that he was still in Lwów during the German occupation and in miserable circumstances. Surviving to see the defeat of Germany, he died in 1945 of lung disease, probably cancer. I had often seen him smoke four or five packs of cigarettes in a day.

In 1929 Kuratowski asked me to participate in a Congress of Mathematicians from the Slavic Countries which was to take place in Warsaw. What sticks in my mind is a reception in the Palace of the Presidium of the Council of Ministers and my timidity at seeing so many great mathematicians, government officials, and important people. This was overcome somewhat when another mathematician, Aronszajn, who was four or five years older than I, said, "Kolego" (this was the way Polish mathematicians addressed each other), "let's go to the other room, the pastry is very good there." (He is now a professor at the University of Kansas in Lawrence.)

The Lwów section of the Polish Mathematical Society held its meetings at the University most Saturday evenings. Usually three or four short papers were given during an hour or so, after which many of the participants repaired to the coffee house to continue the debates. Several times I announced beforehand that I had some results to communicate at one of these sessions when my proof was not complete. I felt confident, but I was also lucky, because I finished the proofs before I had to speak.

I was nineteen or twenty when Stozek asked me to become secretary of the Lwów Section, a job which mainly required sending announcements of meetings and writing up short abstracts of talks for the Society's *Bulletin*. There was, of course, much correspondence between our section and the other sections in Cracow, Poznań, and Wilno. Important problems arose about transferring the administrative seat of the Society from Cracow, the ancient Polish royal city, to Warsaw, the capital, where the headquarters of the Society were eventually located. Needless to say this took a great deal of maneuvering and politicking.

One day a letter came from the Cracow center soliciting the support of the Lwów section. I told Stozek, who was the president of our section, "An important letter just arrived this morning." His reply—"Hide it so no human eye will ever see it again"—was a great shock to my youthful innocence.

The second big congress I attended was held in Wilno in 1931. I went to Wilno by train via Warsaw with Stozek, Nikliborc, and one or two other mathematicians. They kept fortifying themselves with snacks and drinks, but when I pulled out a flask of brandy from my pocket, Stozek burst into laughter and said, "His mama gave it to him in case he should feel faint." This made me acutely aware of how young I was in the eyes of others. For many years I was the youngest among my mathematical friends. It makes me melancholy to realize that I now have become the oldest in most groups of scientists.

Wilno was a marvelous city. Quite different from the cities of the Austrian part of Poland, it gave a definitely oriental impression. The whole city appeared exotic to me and much more primitive than my part of Poland. The streets were still paved with cobblestones. When I prepared to take a bath in my hotel room, the gigantic bathtub had no running water. When I rang the bell a sturdy fellow in Russian boots appeared with three large buckets of hot water to pour into the tub.

I visited the church of St. Ann, the one which Napoleon admired so much on his way to Moscow that he wanted to move it to France.

This was the first and last time I ever visited Wilno. I should mention here that one of the most prominent Polish mathematicians, Antoni Zygmund, was a professor there until World War II. He left via Sweden in 1940 to come to the United States and is now a professor at the University of Chicago.

At the Congress I gave a talk about the results obtained with Mazur on geometrical isometric transformations of Banach spaces, demonstrating that they are linear. Some of the additional remarks we made at the time are still unpublished. In general, the Lwów mathematicians were on the whole somewhat reluctant to publish. Was it a sort of pose or a psychological block? I don't know. It especially affected Banach, Mazur, and myself, but not Kuratowski, for example.

Much of the historical development of mathematics has taken place in specific centers. These centers, large or small, have formed around a single person or a few individuals, and sometimes as a result of the work of a number of people—a group in which mathematical activity flourished. Such a group possesses more than just a community of interests; it has a definite mood and character in both the choice of interests and the method of thought. Epistemologically this may appear strange, since mathematical achievement, whether a new definition or an involved proof of a problem, may seem to be an entirely individual effort, almost like a musical composition. However, the choice of certain areas of interest is frequently the result of a community of interests. Such choices are often influenced by the interplay of questions and answers, which evolves much more naturally from the interplay of several minds. The great nineteenth-century centers such as Göttingen, Paris, and Cambridge (England) all exercised their own peculiar influence on the development of mathematics.

The accomplishments of the mathematicians in Poland between the two world wars constitute an important element in mathematical activity throughout the world and have set the tone of mathematical research in many areas.

This is due in part to the influence of Janiszewski, one

of the organizers of Polish mathematics and a writer on mathematical education, who unfortunately died young. Janiszewski advocated that the new state of Poland specialize in well-defined areas rather than try to work in too many fields. His arguments were, first, that there were not many persons in Poland who could become involved, and second, that it was better to have a number of persons working in the same domain so they could have common interests and could stimulate each other in discussions. On the other hand, this reduced somewhat the scope and breadth of the investigations.

Although Lwów was a remarkable center for mathematics, the number of professors both at the Institute and at the University was extremely limited and their salaries were very small. People like Schauder had to teach in high school in order to supplement a meager income as lecturer or assistant. (Schauder was murdered by the Germans in 1943.) Zbigniew Lomnicki worked as an expert in probability theory in a government institute of statistics and insurance. If I had to name one quality which characterized the development of this school, made up of the mathematicians from the University and the Polytechnic Institute, I would say that it was their preoccupation with the heart of the matter that forms mathematics. By this I mean that if one considers mathematics as resembling a tree, the Lwów group was intent on the study of the roots and the trunk rather than the branches, twigs, and leaves. On a set theoretical and axiomatic basis we examined the nature of a general space, the general meaning of continuity, general sets of points in Euclidean space, general functions of real variables, a general study of the spaces of functions, a general idea of the notions of length, area and volume, that is to say, the concept of measure and the formulation of what should be called probability.

In retrospect it seems somewhat curious that the ideas of algebra were not considered in a similar general setting. It is equally curious that studies of the foundations of physics—in particular a study of space-time—have not been undertaken in such a spirit anywhere to this day.

Lwów had frequent and lively interaction with other mathematical centers, especially Warsaw. From Warsaw Sierpinski would come occasionally, so would Mazurkiewicz, Knaster, and Tarski. In Lwów they would give short talks at the meetings of the mathematical society on Saturday evenings. Sierpinski especially liked the informal Lwów atmosphere, the excursions to inns and taverns, and the gay drinking with Banach, Ruziewicz, and others. (Ruziewicz was murdered by the Germans on July 4, 1941.)

Mazurkiewicz once spent a semester lecturing in Lwów. Like Knaster in topology, he was a master at finding counter examples in analysis, examples showing that a conjecture is not true. His counter examples were sometimes very complicated, but always ingenious and elegant.

Sierpinski, with his steady stream of results in abstract set theory or in set theoretical topology, was always eager to listen to new problems—even minor ones—and to think about them seriously. Often he would send solutions back from Warsaw.

Bronislaw Knaster was tall, bald, very slim, with flashing dark eyes. He and Kuratowski published many papers together. He was really an amateur mathematician, very ingenious at the construction of sets of points and continua with pathological properties. He had studied medicine in Paris during the first World War. Being extremely witty, he used to entertain us with descriptions of the polyglot international group of students and the indescribable language they spoke. He quoted one student he had overheard in a restaurant as having said: "Kolego, pozaluite mnia ein stückele von diesem faschierten poisson," an amalgam of Polish, Russian, Yiddish, German, and French!

Borsuk, more my contemporary, came for a longer visit from Warsaw. We started collaborating from the first. From him I learned about the truly geometric, more visual, almost "palpable" tricks and methods of topology. Our results were published in a number of papers which we sent to Polish journals and to some journals abroad. Actually my first publication in the United States appeared while I was in Lwów. It was a joint paper with Borsuk, published in the *Bulletin of the American Mathematical Society*. We defined the idea of "epsilon homeomorphisms"—approximate homeomorphisms—and the behavior of some topological invariants under such more general transformations—continuous ones, but not necessarily one to one. A joint paper on symmetric products introduced an idea that modifies the definition of a Cartesian product and leads to the construction of some curious manifolds. Some of these might one day find applications in physical theories. They correspond to the new statistics of counting the numbers of particles (not in the familiar classical sense, but rather in the spirit of quantum theory statistics of indistinguishable particles, or of particles obeying the Bose-Einstein or else Fermi-Dirac ways of counting their combinations and dispositions). These cannot be explained here; perhaps this mention will whet the curiosity of some readers.

Kuratowski and Steinhaus, each in a different way, represented elegance, rigor, and intelligence in mathematics. Kuratowski was really a representative of the Warsaw school which flourished almost explosively after 1920. He came to Lwów in 1927, preceded by a reputation for his work in pure set theory and axiomatic topology of general spaces. As editor of *Fundamenta Mathematicae* he organized and gave direction to much of the research in this famous journal. His mathematics was characterized by what I would call a Latin clarity. In the proliferation of mathematical definitions and interests (now even more bewildering than at that time), Kuratowski's measured choice of problems had the quality of what is hard to define—common sense in the abstractions.

Steinhaus was one of the few Polish professors of Jewish descent. He came from a well-known, quite assimilated

Jewish family. A cousin of his had been a great patriot, one of the Pilsudski legionnaires; he was killed during the first World War.

Steinhaus's sense of analysis, his feelings for problems in real variables, in function theory, in orthogonal series manifested a great knowledge of historical development of mathematics and continuity of ideas. Perhaps without so much interest or feeling for the very abstract parts of mathematics, he also steered some new mathematical ideas in the direction of practical applications.

He had a talent for applying mathematical formulations to matters as common as problems of daily life. Certainly his inclinations were to single out problems of geometry that could be treated from a combinatorial point of view—actually anything that presented the visual, palpable challenge of a mathematical treatment.

He had great feeling for linguistics, almost pedantic at times. He would insist on absolutely correct language when treating mathematics or domains of science susceptible to mathematical analysis.

Auerbach was rather short, stooped, and usually walked with his head down. Outwardly timid, he was often capable of very caustic humor. His knowledge of classical mathematics was probably greater than that of most of the other professors. For example, he knew classical algebra very well.

At his instigation Mazur, a few others, and I decided to start a systematic study of Lie groups and other theories which were not strictly in the domain of what is now called Polish mathematics. Auerbach also knew a lot about geometry. I had many discussions with him on the theory of convex bodies, to which Mazur and I contributed several joint papers.

Auerbach and I played chess at the Café Roma and often went through the following little ritual when I began with a certain opening (at that time I did not know any theory of chess openings and played by intuition only). When I made those moves with the king pawn he would say, "Ah! Ruy

Lopez." I would ask, "What is that?" and he would reply, "A Spanish bishop."

Auerbach died during the war. I understand that he and Sternbach took poison while being transported by the Germans to an interrogation session, but I do not know the circumstances of their arrest or anything else about their lives before and during the Nazi occupation.

I believe my collaboration with Schreier started when I was in my second year of studies. Of the mathematicians at the University and at the Polytechnic Institute, he was the only one who was more strictly my contemporary, since he was only six months or a year older and still a student at the University. We met in a seminar room during a lecture by Steinhaus and talked about a problem on which I was working. Almost immediately we found many common interests and began to see each other regularly. A whole series of papers which we wrote jointly came from this collaboration.

We would meet almost every day, occasionally at the coffee house but more often at my house. His home was in Drohobycz, a little town and petroleum center south of Lwów. What a variety of problems and methods we discussed together! Our work, while still inspired by the methods then current in Lwów, branched into new fields: groups of topological transformations, groups of permutations, pure set theory, general algebra. I believe that some of our papers were among the first to show applications to a wider class of mathematical objects of modern set theoretical methods combined with a more algebraic point of view. We started work on the theory of groupoids, as we called them, or semi-groups, as they are called now. Several of these results can be found in the literature by now, but some others have not yet appeared in print anywhere to my knowledge.

Schreier was murdered by the Germans in Drohobycz in April, 1943.

Another mathematician, Mark Kac, four or five years my junior, was a student of Steinhaus. As a beginning undergraduate he had already shown exceptional talent. My connections with Kac developed a little later during my summer visits to Lwów, when I began to spend academic years at Harvard. He also had the good fortune to come to the United States, a few years after I did, and our friendship started in full measure only in this country.

In 1932 I was invited to give a short communication at the International Mathematical Congress in Zürich. This was the first big international meeting I attended, and I felt very proud to have been invited. In contrast to some of the Polish mathematicians I knew, who were terribly impressed by western science, I had confidence in the equal value of Polish mathematics. Actually this confidence extended to my own work. Von Neumann once told my wife, Françoise, that he had never met anyone with as much self-confidence—adding that perhaps it was somewhat justified.

Traveling west, I first joined Kuratowski, Sierpinski, and Knaster in Vienna. They had all come from Kuratowski's summer place near Warsaw; on the way to Zürich the professors decided to stop in Innsbrück. We met some mathematicians from other countries also on their way to the Congress and spent a couple of days there. I remember an excursion by cablecar to a mountain called Hafelekar. This was the first time I was ever above two thousand meters, and the view was beautiful. I remember feeling a little dizzy for a few minutes and identifying this feeling with one I had had previously on several occasions when getting the salient points of proofs of theorems I studied in high school.

The Congress in Zürich was an enormous affair compared to any I had previously attended, but quite small in comparison to those after World War II. I still have a photograph of all the members standing in front of the Technische Hochschule. There for the first time I saw and met many foreign mathematicians.

The meeting was interesting, and I found it stimulating to hear about many types or fields of mathematics other than the ones cultivated in Poland. The diversity of mathematical fields opened new vistas and suggested new ideas to me. In those days I went to almost every available general talk.

Many of the German and West European mathematicians appeared to me nervous; some had facial twitches. On the whole compared to the Poles I knew they seemed less at ease. And even though in Poland there was great admiration for the Göttingen school of mathematics, I again felt, perhaps not justifiably, my own sense of self-confidence.

I gave my own little talk feeling only moderately nervous. The reason for this comparative lack of nervousness, I think, in retrospect, was due to my attitude, compounded of a certain drunkenness with mathematics and a constant preoccupation with it.

Somebody pointed out a short old man. It was Hilbert. I met the old Polish mathematician Dickstein, who was in his nineties and walking around looking for his contemporaries. Dickstein's teacher had been a student of Cauchy in the early nineteenth century, and he still considered Poincaré, who died in 1912, a bright young man. To me this was like going into the prehistory of mathematics and it filled me with a kind of philosophical awe. I met my first American mathematician, Norbert Wiener. Von Neumann was not there, and this was a disappointment. I had heard so much about his visit to Lwów in 1929.

At the hotel swimming pool I met the famous physicist Pauli with Professor Wavre and Ada Halpern. Wavre, Ada's professor, was a Swiss mathematician, known for his studies of the celebrated classical problem of figures of equilibrium of rotating planetary and stellar bodies, among other things. Ada came from Lwów. She was a very good-looking girl who was studying mathematics at the University of Geneva. For a few years I had an off-and-on romance with her. In front of all this company, I turned to Pauli and tried a pun, saying: "This is a Pauli Verbot" (a Paulian physical principle which asserts that two particles with the same characteristics cannot occupy the same place), referring to Wavre and me who were both there in the company of this pretty young lady.

Another interesting encounter occurred one afternoon in the woods around the famous Dolder Hotel. Having lost my way, I ran into Paul Alexandroff and Emmy Noether walking together and discussing mathematics. Alexandroff knew about some of my work for I had sent him reprints and we had had some previous mathematical correspondence. In fact one of the great joys of my life had been to receive a letter from him addressed to Professor S. Ulam. During this encounter he suddenly said to me: "Ulam, would you like to come to Russia? I could arrange everything and would like very much to have you." As a Pole, and with my rather capitalistic family background, his invitation flattered me, but such a trip appeared quite unthinkable.

The Congress over, after a little excursion to Montreux with Kuratowski and Knaster I returned to Poland in time to take my Master's degree.

I had an almost pathological aversion to examinations. For over two years I had neglected to take the examinations which were usually necessary to progress from one year to the next. My professors had been tolerant, knowing that I was writing original papers. Finally, I had to take them—all at once.

I studied for a few months, took a kind of comprehensive examination and wrote my Master's thesis on a subject which I thought up myself. I worked for a week on the thesis, then wrote it up in one night, from about ten in the evening until four in the morning, on my father's long sheets of legal paper. I still have the original manuscript. (It is unpublished to this day.) The paper contains general ideas on the operations of products of sets, and some of it outlines what is now called Category Theory. It also contains some individual results treating very abstractly the idea of a general theory of many variables in diverse parts of mathematics. All this was in the fall of 1932 upon my return from Zürich.

In 1933 I took my Doctor's examination. The thesis was published by Ossolineum, an establishment which printed the Lwów periodical *Studia Mathematica*. It combined several of my earlier papers, theorems, and generalizations in measure theory.

My degree was the first doctorate awarded at the Polytechnic Institute in Lwów from the new Department of General Studies which had been established in 1927. It was the only department that gave Master's and Doctor's degrees, all the others being engineering degrees.

The ceremony was a rather formal affair. It took place in a large Institute hall with family and friends attending. I had to wear a white tie and gloves. My sponsors Stozek and Kuratowski each gave a little speech describing my work and the papers I had written. After a few words about the thesis, they handed me a parchment document.

The "aula"—the large hall in which the ceremony took place—was decorated with traditional frescoes. These were very much like some I saw twenty years later on the walls of the MIT cafeteria. The MIT frescoes depict scantily dressed women in postures of flight, symbolizing sciences and arts, and a large female figure of a goddess hovering over a recoiling old man. I used to joke that it represented the Air Force giving a contract to physicists and mathematicians. In Fuld Hall, the Institute Building in Princeton, there is also an old painting in the tea room where people assemble for conversation in the afternoon. There again one sees an old man who seems to be shying away from an angel coming down from the clouds. When I was told that nobody knew what it was supposed to represent, I suggested that it might be a representation of Minna Ries, the lady mathematician who directed the Office of Naval Research at the time, proposing a Navy contract to Einstein, who is recoiling in horror.

After the examinations and ceremonies I published a few more papers and then had to take it easy for the rest of 1933, for a bad paratyphoid infection left me weak for several months—one of the rare times in my life when I was seriously ill.

But not all was serious work and no play. In the early 1930s, a high school teacher of science by the name of Hirniak, a wizened, small man, came to our coffee house. He would sit a few tables away from us, sipping vodka and coffee in turn, and busily scribbling on a pad of paper. Every once in a while he would get up and join our table to gossip or kibitz when Nikliborc and Stozek played chess. Nikliborc would repeat with glee: "Gehirn [brain in German], Gehirniak!"

Hirniak, who taught mathematics, physics, and chemistry, was trying to solve Fermat's famous problem. This is one of the best-known unsolved problems in mathematics, and for a long time has attracted cranks as well as amateurs, who regularly produce false or very incomplete proofs of Fermat's conjecture.

Hirniak was a fixture at the coffee house, and his conversation was delightfully picturesque and full of unconsciously humorous statements. We would collect and repeat them to each other; I used to paste some of them on the walls of my room at home.

It turned out that my father knew Hirniak, whose wife owned a large soda-water factory and whose legal affairs were handled by my father's firm. My father considered Hirniak a humorously foolish person. When he saw my collection of Hirniak maxims, I believe he was surprised and perhaps even wondered about my sanity. I had to explain to him the subtlety of the humor and its special appeal to mathematicians.

Hirniak would tell Banach, for instance, that there were still some gaps in his proof of Fermat's problem. Then he would add, "The bigger my proof, the smaller the hole. The longer and larger the proof, the smaller the hole." To a mathematician this constitutes an amusing formulation. He would make weird statements about physics. For example, he would say that half the elements in the periodic table are metals and the other half are not. When someone pointed out that this was not quite correct, he would reply: "Ah, but by definition we can call a few more of them metals!" He had a wonderful way of taking liberties with definitions.

He studied in Göttingen and described how he would drink cups of wine from an automatic dispenser. Once something went wrong in the machine, and the wine continued to flow. Hirniak continued to drink until he found himself lying on the ground surrounded by a group of people. He heard someone ask, "Vielleicht ist etwas los?" (Maybe something went wrong?). He replied, "Vielleicht nicht." (Maybe not.) At this, he was carried home in triumph on the shoulders of the crowd.

Here is the story that entertained von Neumann so much when I told him about Hirniak years later in Princeton: one day Hirniak told Banach, Mazur, and me that he had almost proved Fermat's conjecture and that American reporters would find out about it and would come to Lwów and say: "Where is this genius? Give him one hundred thousand dollars!" And Banach would echo: "Give it to him!" After the war, Johnny said to me one day in Los Alamos, "Remember how we used to laugh at Hirniak's hundred thousand dollars story? Well, he was right, he was the real prophet while we were laughing like fools." What Johnny was referring to, of course, was that representatives of the Defense Department, the Air Force, and the Navy were traveling around the country at the time bountifully dispensing research contracts to scientists. The average contract amounted to about one hundred thousand dollars. "Not only was he right," said Johnny, "but he even foresaw the correct amount!"

Sometime around 1933 or 1934, Banach brought into the Scottish Café a large notebook so that we could write statements of new problems and some of the results of our discussions in more durable form. This book was kept there permanently. A waiter would bring it on demand and we would write down problems and comments, after which the waiter would ceremoniously take it back to its secret cache. This notebook was later to become famous as "The Scottish Book."

Many of the problems date from before 1935. They were discussed a great deal by those whose names were included. Most of the questions posed were supposed to have received considerable attention before an "official" inclusion could be considered. In several cases, the problems were solved on the spot and the answers included.

The city of Lwów and the Scottish Book were fated to have a very stormy history within a few years of the book's inception. After the outbreak of World War II, the city was occupied by the Russians. From items toward the end of the book it is evident that some Russian mathematicians must have visited the town. They left several problems and offers of prizes for their solution. The last date appearing in the book is May 31, 1941. Item No. 193 contains a rather cryptic set of numerical results signed by Steinhaus dealing with the distribution of the number of matches in a box! After the start of the war between Germany and Russia, the city was occupied by German troops in the summer of 1941, and the notes ceased. The fate of the book during the remaining years of the war is not known to me. According to Steinhaus, this document was brought to Wroclaw (formerly Breslau) by Banach's son, now a neurosurgeon in Poland.

During my last visit to Lwów in the summer of 1939, a few days before I left I had a conversation with Mazur on the likelihood of war. People were expecting another crisis like Munich and were not prepared for an imminent world war. Mazur said to me, "A world war may break out. What shall we do with the Scottish Book and our joint unpublished papers? You are leaving for the United States and presumably will be safe. In case of a bombardment of the city, I shall put the manuscripts and the book in a case, which I shall bury in the ground." We even decided on a location. It was to be near the goal post of a football field just outside the city. I do not know whether any of this really happened, but apparently the manuscript of the Scottish Book survived in good shape, for Steinhaus sent me a copy of it after the war. I translated it in 1957 and distributed it to many mathematical friends in the United States and abroad.

Of the surviving mathematicians from Lwów many are continuing their work today in Wroclaw. The tradition of the Scottish Book continues. Since 1945 new problems have been posed and recorded and a new volume is in progress.

CHAPTER 3

Travels Abroad

1934

BY 1934 I had become a mathematician rather than an electrical engineer. It was not so much that I was doing mathematics, but rather that mathematics had taken possession of me. Perhaps this is a good place to stop for a moment and ponder what being a mathematician means.

The world of mathematics is a creation of the brain and can be visualized without external help. Mathematicians are able to work on their subject without any of the equipment or props needed by other scientists. Physicists (even theoretical physicists), biologists, and chemists need laboratories—but mathematicians can work without chalk or pencil and paper, and they can continue to think while walking, eating, even talking. This may explain why so many mathematicians appear turned inward or preoccupied while performing other activities. This is quite pronounced and quantitatively different from the behavior of scientists in other fields. Of course, it depends on the individual. Some, like Paul Erdös, have this characteristic in the extreme. His preoccupation with mathematical construction or reasoning occupies a very large percentage of his waking hours, to the exclusion of everything else.

As for myself, ever since I started learning mathematics I would say that I have spent—regardless of any other activity—on the average two to three hours a day thinking and two to three hours reading or conversing about mathematics. Sometimes when I was twenty-three I would think about the same problem with incredible intensity for several hours without using paper or pencil. (By the way, this is infinitely more strenuous than making calculations with symbols to look at and manipulate.)

On the whole, I still find conversation with or listening to other people an easier and pleasanter way of learning than reading. To this day I cannot read "how to" instructions in printed form. Psychologically, these are indigestible for me.

Some people prefer to learn languages by the rules of grammar rather than by ear. This can be said to be true of mathematics—some learn it by "grammar" and others "from the air." I learned my mathematics from the air.

For example, I learned, subconsciously, from Mazur how to control my inborn optimism and how to verify details. I learned to go more slowly over intermediate steps with a skeptical mind and not to let myself be carried away. Temperament, general character, and "hormonal" factors must play a very important role in what is considered to be a purely "mental" activity. "Nervous" characteristics play an enormous role in one's intellectual development. By the age of about twenty, when development is supposed to be fully completed, some of these acquired traits are perhaps essentially frozen and have become a permanent part of our makeup.

Mathematics is supposed to be in essence only a very general and precise language, but perhaps this is only partially true. There are many ways of expressing oneself. A person who starts early has some particular way of organizing his memory or devises his own particular system for arranging impressions. A "subconscious brewing" (or pondering) sometimes produces better results than forced, systematic thinking, as when planning an overall program in contrast to pursuing a specific line of reasoning. Forcing oneself to persist in a logical exploration becomes a habit, after which it ceases to be forcing since it comes automatically (as a subroutine, as computer people like to say). Also, even if one cannot define what we call originality, it might to some extent consist of a methodical way of exploring avenues—an almost automatic sorting of attempts, a certain percentage of which will be successful.

I always preferred to try to imagine new possibilities rather than merely to follow specific lines of reasoning or make concrete calculations. Some mathematicians have this trait to a greater extent than others. But imagining new possibilities is more trying than pursuing mathematical calculations and cannot be continued for too long a time.

An individual's output is, of course, conditioned by what he can accomplish most easily and this perhaps restricts its scope. In myself I notice a habit of twisting a problem around, seeking the point where the difficulty may lie. Most mathematicians begin to worry when there are no more difficulties or obstacles for "new troubles." Needless to say, some do it more imaginatively than others. Paul Erdös concentrates all the time, but usually on lines which are already begun or which are connected to what he was thinking about earlier. He doesn't wipe his memory clean like a tape recorder to start something new.

Banach used to say, "Hope is the mother of fools," a Polish proverb. Nevertheless, it is good to be hopeful and believe that with luck one will succeed. If one insists only on complete solutions to problems, this is less rewarding than repeated tries which result in partial answers or at least in some experience. It is analogous to exploring an unknown country where one does not immediately have to reach the end of the trail or all the summits to discover new realms.

It is most important in creative science not to give up. If you are an optimist you will be willing to "try" more than if you are a pessimist. It is the same in games like chess. A really good chess player tends to believe (sometimes mistakenly) that he holds a better position than his opponent. This, of course, helps to keep the game moving and does not increase the fatigue that self-doubt engenders. Physical and mental stamina are of crucial importance in chess and also in creative scientific work. It is easier to avoid mistakes in the latter, in that one can come back to rethinking; in chess one is not allowed to reconsider moves once they have been made.

The ability to concentrate and the decrease in awareness of one's surroundings come more naturally to the young. Mathematicians can start very young, in some cases in their teens. In Europe, even more than in America, mathematicians exhibit precocity, education in European high schools having been several years ahead of the more theoretical education in the United States. It is not unusual for mathematicians to achieve their best results at an early age. There are some exceptions; for instance, Weierstrass, who was a high-school teacher, achieved his best results when he was forty. More recently, Norman Levinson proved a very beautiful theorem when he was sixty-one or sixty-two.

At twenty-five, I had established some results in measure theory which soon became well known. These solved certain set theoretical problems attacked earlier by Hausdorff, Banach, Kuratowski, and others. These measure problems again became significant years later in connection with the work of Gödel and more recently with that of Paul Cohen. I was also working in topology, group theory, and probability theory. From the beginning I did not become too specialized. Although I was doing a lot of mathematics, I never really considered myself as only a mathematician. This may be one reason why in later life I became involved in other sciences.

In 1934, the international situation was becoming ominous. Hitler had come to power in Germany. His influence was felt indirectly in Poland. There were increasing displays of inflamed nationalism, extreme rightist outbreaks and anti-Semitic demonstrations.

I did not consciously recognize these portents of things to come, but felt vaguely that if I was going to earn a living by myself and not continue indefinitely to be supported by my father, I must go abroad. For years my uncle Karol Auerbach had been telling me: "Learn foreign languages!" Another uncle, Michael Ulam, an architect, urged me to try a career abroad. For myself, unconscious as I was of the realities of the situation in Europe, I was prompted to arrange a longish trip abroad mainly by an urge to meet other mathematicians, to discuss problems with them and, in my extreme self-confidence, try to impress the world with some new results. My parents were willing to finance the trip.

My plans were to go west (go west, young man!); first I wanted to spend a few weeks in Vienna to see Karl Menger, a famous geometer and topologist, whom I had met in Poland through Kuratowski. This was the fall of 1934, right after the assassination of the Austrian Premier Dollfuss. Vienna was in a state of upheaval, but I was so absorbed and almost perpetually drunk with mathematics that I was not really aware of it.

After a couple of days in a Vienna hotel, I moved to a private boarding house near the University, where a widowed lady rented rooms to students. This was quite a common arrangement in those days. The house was on a little street named after Boltzmann, a great physicist of the nineteenth century, one of the principal creators of the kinetic theory of gases and of thermodynamics.

I visited Menger and at his house met a brilliant young Spanish topologist named Jimenez y Flores, who had already some nice results to his credit. We talked mathematics a good deal. He seemed very well known in night clubs and introduced me to the life of a young man about town.

From Vienna I traveled to Zürich to meet Heinz Hopf, the topologist. He was a professor at the famous Technische Hochschule, with whom I had corresponded. Hopf knew something about my topological results and invited me to visit the institute to give two lectures. One was about work I had done jointly with Borsuk on the "antipodal theorem," a topological problem. I spoke in German, in a lecture room of the department of agriculture. I recall there were many pictures of prize cows along the walls, which seemed to look at me with sadness and commiseration.

Nevertheless, this visit to Zürich was quite fruitful. I also met a physicist named Grossman, who was a few years my senior and widely traveled. He recommended hotels in France and in England to suit my purse. We discussed philosophy and the role of mathematics in physics.

After two weeks in Zürich, I went to Paris for five weeks, and that was sheer delight. I had been in France before, but this was my first visit to Paris. My uncle Michael's wife happened to live there at the time and she kindly offered to receive me and to send to my modest hotel her chauffeured limousine to take me sightseeing. I was so embarrassed at the thought of being seen arriving in a Rolls Royce or a Dusenberg at the Louvre or some other museum, it felt so incongruous, that I declined her offer.

I went to the Institut Poincaré with a letter of introduction from one of my professors to the famous old mathematician Elie Cartan. When I entered his office I plunged directly into a mathematical discussion, telling him how I had an idea for a simple and general proof for solving Hilbert's fifth problem on continuous groups. At first he said he did not quite follow my reasoning, but then he added, "Ah! I see now what you want to do." Cartan's little white goatee, vivacious smile, and sparkling eyes gave him an appearance which somehow fitted my mental image of all French mathematicians. He was remarkable for many reasons, not the

least because he had done some of his best work in his fifties, when the creativity of most mathematicians is on the
decline.

I attended several seminars and talks at the Institut Poincaré and at the Sorbonne. At the first seminar, a young
Frenchman named De Possel happened to be talking about
one of my results. It made me swell with pride. (De Possel
is still teaching in Paris.) I was invited to give a talk in a
salle named after the mathematician Hermite, another in the
salle Darboux. These halls and streets such as Rue Laplace,
Rue Monge, Rue Euler, visible signs that the abstractions
worked on by mathematicians were somehow appreciated,
became heady wine and added to my general state of eu-
phoria. In my youthful way I wondered, "If only some day a
hundred years from now, a little street or even an alley
could be named after me."

In October I decided to go to Cambridge, England.
Steinhaus had given me a letter of introduction to Professor
G. H. Hardy, a legendary figure in mathematics. In Lwów
his discoveries in the theory of numbers were well known,
and my friend Schreier used to present his papers in seminars. Stories about Hardy's eccentricities were widely told.

I found that belonging to the upper middle class did
often facilitate things in England. In Dover, when, by mistake, I left the boat through the wrong door, two British
plainsclothesmen intercepted me and wanted to know
where I was going. I must have looked younger than my
twenty-five years for one of them asked me what my father's
occupation was. When I replied that he was a barrister, the
man turned to his partner and said in a typically British way:
"He is all right, his father is a barrister." I thought it was
very comical that they took my word so easily for this piece
of information.

After a few hours in London, I took an evening train for
Cambridge. The train stopped every few minutes at stations,
all in the dark, whose names were not visible. I asked a
young man in my compartment: "How can you tell when it

is Cambridge?" He thought for a moment and replied: "I am afraid you can't." After another silence, I tried to start a new conversation by asking him what he thought about the political situation and whether he thought England would intervene in the Ruhr and help France. He pondered again for a minute or two and answered: "I am afraid not!" I was absolutely delighted by what seemed to me such very, very British utterances. As my knowledge of British mores derived mainly from Dorothy Sayers and Agatha Christie novels, somehow this fitted in.

I got off at Cambridge and went to a hotel called the Garden House which had been recommended by Grossman in Zürich. Since my father was financing my travels, each week I received five or ten pounds at Barclay's bank from my uncle's bank in Lwów. In those days this was almost affluence. I walked around Cambridge, admiring the University buildings and looking into bookstores. (I already had a pronounced book-buying—or, at least, book-handling—mania.) The Sherlock Holmes and Conan Doyle atmosphere I saw in many places enchanted me.

I hunted up a few mathematicians. Besicovitch, a Russian émigré from the Russian Revolution, was one with whom I had corresponded. He had solved one of my problems which had appeared in *Fundamenta* and had published a paper on it. It was really the first non-obvious example of an "ergodic transformation," a mapping of a plane onto itself, in which the successive images of a point were dense in the whole plane.

Besicovitch invited me to visit him in his rooms in Trinity College. When I entered his place, he said nonchalantly, "Newton lived here, you know." This gave me such a shock that I almost fainted. Landmarks in the great history of science like this literally kept me in a state of excitement for the rest of my stay in England.

Besicovitch and I talked mathematics. I wonder if many older persons were accustomed to such young men coming into their rooms and abruptly plunging into scientific problems and theorems without even explaining their own presence or exchanging greetings first. My friend Erdös is still like that at the age of sixty. Von Neumann too, who was so urbane and interested in politics and gossip, would often shift abruptly from a general conversation to technical scientific remarks.

In several ways, my stay in Cambridge was one of the most pleasant periods of my life—intellectually and in a psychological sense. Besicovitch invited me to a dinner at High Table at Trinity College. This dinner was one of the high points of my entire life until then. Present were G. H. Hardy, J. J. Thomson, Arthur S. Eddington, and other famous scientists, and there I was, sitting only a few feet away. The conversation was exciting. I listened to every word. We sat under an old portrait of Henry VIII. Food was served in ancient silver dishes. I noticed that Besicovitch ate with an excellent appetite. After dinner we moved to another room, and he drank brandy after brandy, while the others cast furtive but admiring glances in his direction.

Hardy told anecdotes, one of which I remember. As a youth he was once walking through a thick fog with a man of the cloth and they saw a boy with a string and a stick. Hardy's clergyman compared this to the invisible presence of God which can be felt but not seen. "You see, you cannot see the kite flying, but you feel the pull on the string." Hardy knew, however, that in a fog there is no wind and so kites cannot fly. Hardy believed that, in mathematics, the Cambridge examinations called "triposes" were nonsensical. As a demonstration, he persuaded George Polya (who, if anything, was a master of computation and manipulation in classical analysis) to take the mathematics tripos without previous coaching. Polya supposedly failed miserably.

I met Subrahmanyan Chandrasekhar, a brilliant young astrophysicist from India. We had a few meals together at Trinity, where he was a fellow. He collaborated with Eddington for whom he had mixed feelings of admiration and rivalry. A year later, the vacancy in the Society of Fellows at

Harvard which I was invited to fill resulted from Chandra-sekhar's acceptance of an assistant professorship in Chicago.

We met again much later when he was a consultant in Los Alamos working on the theory of turbulence and other hydrodynamical problems. Chandra, as he is known among his friends, is one of the world's most brilliant and prolific mathematical astronomers. His books are classics in his field.

During this stay in Cambridge, Michaelmas term 1934, the university or the authorities of the individual colleges for women—Girton and Newnham—abolished the old rule which forbade men lecturers on the college premises. I was invited to give a seminar on topology. I was, if I am not mistaken, the first male in the history of Girton to cross its threshold to give a lecture.

Of all the scientists I had known in Poland, the only one I saw while in Cambridge was Leopold Infeld, who was a docent in Lwów. I knew him from our coffee houses, and we saw each other a few times in Cambridge.

Infeld was tall, well over six feet, quite portly, with a large head and a large face. He was Jewish, from a simple orthodox background. In his autobiography he devoted much space to a description of his fight to achieve an education and an academic position, neither of which was easily attained.

He was rather gay and witty. I remember what seemed to me a bright remark he made after a month's stay in England about the difference between Polish and English "intellectual" conversations. He said that in Poland people talked foolishly about important things, and in England intelligently about foolish or trivial things.

Infeld was a very ambitious man and had a colorful career. I do not think his talent for physics or mathematics was quite up to his ambitions. In Poland, I had had some doubts about his real understanding of the mathematics of the deeper parts of general relativity. Perhaps it was because of his rather limited background in fundamental mathematics.

The popular articles he wrote in one of the Warsaw newspapers were well written but, it seemed to me, not always mathematically exact. At the time my sights were set very high and I expected even newspaper articles about science to be comparable to Poincaré's wonderful writings on popular science or Eddington's explanations of relativity theory for popular audiences.

Infeld came to Princeton later, a few weeks after I did, and collaborated with Einstein on the well-known Einstein-Infeld book on physics, which became a best seller. He had met Einstein in Berlin, and in his autobiography he describes how impressed he had been by his friendliness and ability to put people at ease. In Princeton I hardly saw him; he was not part of the von Neumann crowd.

The Cambridge architecture, the medieval buildings, the beautiful courtyards, the walks I took through the town, some with L. C. Young (now a professor at the University of Wisconsin), are still among the strongest visual impressions of my life. Like my walks through the Paris of the French Revolution, these have somehow influenced my tastes, associations, readings, and studies to this day.

Early in 1935, I returned from Cambridge to Poland. It was now time to think seriously about a university career, although those were difficult times in which to find even a modest "docent" position. A series of accidental letters was to change this; in one of them, luckily for me, I received an invitation to visit the United States.

PART II

A Working
Mathematician
in America

CHAPTER 4

Princeton Days

1935–1936

I FIRST heard about John von Neumann from my high school teacher Zawirski. Kuratowski also described von Neumann's results and his personality. He told me how in a Berlin taxicab von Neumann had explained in a few sentences much more than he, Kuratowski, would have gotten by correspondence or conversation with other mathematicians about questions of set theory, measure theory, and real variables. Banach, too, talked about him. He told me how at the 1927 Lwów meeting he and other mathematicians, Stozek among them, had made von Neumann drunk at the congress banquet by plying him with vodka, to the extent that he had to leave the table to go to the toilet. He came back and continued the mathematical conversation without having lost his train of thought.

It was only toward the end of 1934 that I entered into correspondence with von Neumann. He was then in the United States, a very young professor at the Institute for Advanced Studies in Princeton. I wrote him about some problems in measure theory. He had heard about me from Bochner, and in his reply he invited me to come to Princeton for a few months, saying that the Institute could offer me a $300 stipend. I met him shortly after my return from England.

In the fall of 1935 a topology conference had been organized in Moscow. Alexandroff invited me to attend. At that time relations between Poland and Soviet Russia were strained. Passport applications in Poland for travel to Russia involved so much red tape that I did not receive my passport in time and thus missed going to the meeting. Von Neumann wrote me that he would be passing through Warsaw on his way back from Moscow, suggesting that we meet there. Samuel Eilenberg, a young Warsaw mathematician well known for his very ingenious topological results, and I went to meet the returning western group. At the station von Neumann (whom I was seeing for the first time) was accompanied by two American mathematicians, Garrett Birkhoff and Marshall Stone. We all conversed in English. Eilenberg spoke it a little; I spoke it adequately thanks to my Cambridge stay. Von Neumann would break into German occasionally.

From Kuratowski's description, I had imagined him to be slim, as he apparently had been in 1927. He was instead rather plump, though not as corpulent as he was to become later. The first thing that struck me about him were his eyes—brown, large, vivacious, and full of expression. His head was impressively large. He had a sort of waddling walk. (This reminds me that when I first saw his grandson, Malcolm, the son of his daughter Marina, I found it uncanny to watch this little three-year-old perambulate down a long hotel corridor with his grandfather's waddle, holding his hands behind his back exactly like Johnny. Since he had been born after his grandfather's death, he could not possibly have been imitating him. It would seem that gestures, motions, and other time-dependent phenomena—not merely static characteristics or matters of spatial configuration—can be transmitted genetically.)

Von Neumann appeared quite young to me, although he was in his early thirties, some five or six years older than I. (I have always had mixed feelings toward people older than myself: on the one hand, something like respect; on the other, a slight feeling of superiority, of having a greater share in the future.) At once I found him congenial. His habit of intermingling funny remarks, jokes, and paradoxical anecdotes or observations of people into his conversation, made him far from remote or forbidding.

During this brief visit, Stone, von Neumann, and Birkhoff gave a joint seminar at the Warsaw section of the Polish Mathematical Society. Their subject was the lattice theoretical foundations of quantum theory logic. Von Neumann gave most of the lecture, Birkhoff talked briefly, and Stone asked questions. Actually I had mixed impressions about this talk. I was not at all convinced that it had much to do with novel physical ideas. In fact, I thought the points were a bit stretched and the big notion of quantum theory logic a bit artificially contrived. I had a number of other conversations with Johnny, mainly on measure theory (about which I had sent him reprints of my earlier papers). We also talked a little about his recent work on the theory of Hilbert space operators, although I was not especially knowledgeable about or interested in it. Then he gave me some practical advice about my forthcoming trip to Princeton.

In connection with the Moscow topological meeting, several years after World War II, I received a letter from the French mathematician Leray, who with the Lwów mathematician Juliusz Schauder had written a celebrated paper on fixed points for transformations in function spaces and applications in the theory of differential equations. Schauder, our mutual friend, was murdered by the Nazis. Leray wanted to have a photograph of him for himself and for Schauder's daughter who survived the war and lives in Italy. But he could not find any in Poland or anywhere and he wrote me asking whether I might have a snapshot. Some months after Johnny von Neumann's death I was looking at some of the books in his library and a group photo of the participants in the Moscow conference fell out. Schauder was there, as were Alexandroff, Lefschetz, Borsuk, and some dozen other topologists. I sent this photograph to Leray. It has since been reproduced in several publications.

Just as in Lwów, the Warsaw mathematicians gathered in a pastry shop and discussed mathematics for hours. They also frequented the famous Fuker wine shop in the old town. This is where Eilenberg and I took Johnny and his companions to drink Fuker's celebrated hydromel. Here he entertained us with the story of how at the request of Princeton friends he had bought several pounds of caviar in Moscow to bring back to the U.S. and had asked a steward to store it in the icebox of the restaurant car. In the morning when they woke up, in Poland, they discovered that the restaurant car had been uncoupled at the Polish-Russian border. They were returning to the States caviarless! He talked also about his own decision to emigrate to America and the general impracticality and lack of foresight of European scientists. In the German universities the number of existing and prospective vacancies for professorships was extremely small—something like two or three in the entire country for the next two years. Yet most of the two or three score docents counted on obtaining a professorship in the near future. With his typical rational approach, von Neumann computed that the expected number of professorial appointments within three years was three, whereas the number of docents was forty. This is what had made him decide to emigrate, not to mention the worsening political situation, which made him feel that unhampered intellectual pursuits would become difficult. In 1930, he accepted an offer of a visiting professorship at Princeton University, and, in 1933, shortly after the creation of the Institute, he was invited to become the youngest member of the permanent faculty of the Institute for Advanced Studies.

In December 1935 I sailed on the English ship *Aquitania* from Le Havre on my first transatlantic crossing. The weather was beautiful for the first two days; then a violent storm sprang up, and I became seasick. When the boat approached New York, the sea calmed down and my seasickness stopped.

After two days in New York, I tried to reach von Neumann in Princeton but got no answer, so I called the Institute. It was quite an experience for me to go into an American telephone booth for the first time. When the operator said, "Hold the wire!" I did not understand the expression and asked "Which wire should I hold?" I reached Solomon Lefschetz, a professor at the University, who told me how to get to Princeton from New York. He said it was very easy, that there were trains every hour. I could not understand that. Princeton I knew was a very small town, why should there be a train every hour? I did not know it was on the main line to Philadelphia and Washington.

In Princeton I went straight to register at the Institute which was housed in Fine Hall, a University building, for it did not have its own quarters yet. A young and pretty Miss Flemming and an older Miss Blake received me. I was greeted with smiles. It surprised me, and I wondered if there was something a little funny about the way I was dressed or whether my trousers were not properly buttoned (there were no zippers in those days).

I checked in at a boarding house and went directly to visit von Neumann in his large and impressive house. A black servant let me in, and there was Solomon Bochner in the living-room and a baby crawling on the floor. (The baby was von Neumann's daughter Marina, six months old at the time.) Marietta, his first wife, also a Hungarian, greeted me. I knew of Bochner since we had corresponded about mathematics. Bochner and von Neumann were talking politics.

Von Neumann expressed great pessimism about the possibility of a war in Europe. (This was about three years before the actual outbreak.) Apparently he had a rather clear picture of the catastrophes to come. He saw Russia as the chief antagonist to Nazi Germany. Believing that the French army was strong, I asked, "What about France?" "Oh! France won't matter," he replied. It was really very prophetic.

My lodgings were in a boarding house on Vandevanter Street, if I remember correctly. There were six or eight other men there, not all students, and we ate together. I remember how the conversation was at first completely incomprehensible to me although I knew English. The American accent took me by surprise, and I missed most of what was being said. Then after a week I understood everything. This is a common experience, not only with languages but also with mathematics—a discontinuous process. Nothing, nothing, at first, and suddenly one gets the hang of it.

I became a frequent visitor at the von Neumanns who were very sociable and held parties two or three times a week. These were not completely carefree; the shadow of coming world events pervaded the social atmosphere. There I met the Alexanders, who were great friends of the von Neumanns. James Alexander, also a professor at the Institute, was an original topologist, the creator of novel problems and strange "pathological" examples of topological objects. He was the scion of a wealthy family and very eccentric.

At a party a few weeks later, I saw a man who must have been fifty but seemed to me infinitely old—I was twenty-six at the time. He was sitting in a big chair with a nice young lady on his knee. They were drinking champagne. I passed by Johnny and asked, "Who is this gentleman?" "Oh! don't you know? He is von Kármán, the famous aerodynamicist." Von Kármán was one of Johnny's friends. And he added: "Don't you know that he invented consulting?" Von Kármán was one of the first scientists who learned to fly a plane in the first World War. He told me that he had one of the low numbered international flying licenses. His flying experiences directly influenced his ideas about jet engines, which became so important in World War II developments. Much later I came to know him quite well. He used to say that engineers are people who perpetuate the mistakes made in the previous generation. In 1968 I found myself with him in Israel at a meeting on hydrodynamics. By then he was a rich old man. It was his first visit to that country, and he was so moved and impressed by what he saw that he gave away five- and ten-dollar bills as tips to waiters and taxi drivers, no matter what the size of the check.

Johnny was always impressed by people who were successful in political or organizational activities or in physical exploits, and he cultivated them. It was Johnny who said as we walked past the elegant Gothic Princeton University Chapel: "This is our one-million-dollar protest against materialism." I do not know whether this bon mot was his own or not, but it shows his sense of humor about money.

In those days he still called me "Mr. Ulam." Once he said to me as we were driving in the rain and were caught in a traffic jam, "Mr. Ulam, cars are no good for transportation anymore, but they make marvelous umbrellas." I often remember this when I am caught in the traffic jams of today. Johnny always loved cars but he drove somewhat carelessly.

Johnny lived rather sumptuously. The professors at the Institute were the highest paid academics in the United States —paid even more than at Harvard. This tended to create animosities between Institute and University professors. Also their compensation was in sharp contrast to the almost negligible stipends offered to the fellows and visitors of the Institute.

The great name, the great celebrity, the great light, so far as the general public was concerned, was, of course, Albert Einstein. I first met his assistant, Mayer, a mathematician and a strange person. Then I was introduced to Einstein himself and noticed his rather peculiar English. He would say: "He is a very good formula," pointing at something on the blackboard!

A cousin of mine, Andrzej Ulam, a banker, came to New York on business about two months after my arrival, and I invited him to visit me in Princeton. It happened that during that week I was giving a talk in some seminar, and my name was listed on the same page of the Institute's Bulletin as the announcement of Einstein's regular weekly seminar. This impressed him enormously; he mentioned it in a letter home, and my reputation among friends and family in Poland was made.

Hermann Weyl was also a professor at the Institute. I met him in Princeton and went to his home several times. He was a legendary figure, much older than von Neumann. He had the same wide breadth of interests which impresses me so much. My friend Gian-Carlo Rota, now a professor at MIT, told me much later that he had heard Weyl's original symmetry lectures and was enormously impressed. They were a bit heavy, but at the same time they gave a feeling of universal culture. More recently Weyl's purely mathematical schemata or algebraic entities found essential applications as models for the properties of the mysterious neutrino particles and so-called weak interactions important in the beta decay of nuclei.

After Weyl's first wife died he remarried and lived in Switzerland for a time. He was unaware of the rules and regulations governing the length of time a naturalized American citizen may live abroad without returning to this country and still retain his American citizenship. He lost his by negligence. When it happened, everyone was shocked. Members of the Mathematical Society and the National Academy of Sciences wanted to have him reinstated as a U.S. citizen. This required a special bill in Congress, and some friends asked me to intervene with Senator Anderson, whom I knew well, to help in this matter. In the meantime Weyl had collapsed in a street in Zürich while putting a letter in a mailbox and died of a heart attack.

Every day I went to the Institute for five or six hours. By that time I had quite a number of published papers, and people there knew some of them. I talked with Bochner a good deal and soon after my arrival communicated to him a problem about "Inverting the Bernouilli law of large numbers." Bochner proved the theorem and published it in *The Annals of Mathematics*. (By the way, this problem is still solved in a simple case only. The inverse of the law of large numbers, those requiring a measure in a space of measures, has not yet been proved.)

I went to lectures and seminars, heard Morse, Veblen, Alexander, Einstein, and others, but was surprised how little people talked to each other compared to the endless hours in the coffee houses in Lwów. There the mathematicians were genuinely interested in each other's work, they understood one another because their work revolved around the central theme of set theoretical mathematics. Here, in contrast, several small groups were working in separate areas, and I was somewhat disappointed at this lack of curiosity even though the Institute and the University had a veritable galaxy of celebrities, possibly constituting one of the greatest concentrations of brains in mathematics and physics ever to be assembled. Being a malicious young man, I told Johnny that this reminded me of the division of rackets among Chicago gangsters. The "topology racket" was probably worth five million dollars; the "calculus of variations racket," another five. Johnny laughed and added, "No! That is worth only one million."

There was another way in which the Princeton atmosphere was entirely different from what I expected: it was fast becoming a way station for displaced European scientists. In addition, these were still depression days and the situation in universities in general and in mathematics in particular was very bad. People with impressive backgrounds and good credentials (not only visitors like me, but native-born Americans) were still without jobs several years after getting their doctorates. A very able mathematician and logician friend of mine, who is now a member of the National Academy of Sciences, was then on a miserable stipend at Princeton waiting for a position to open somewhere. One day a telegram came offering him an instructorship at twelve hundred dollars a year. He told me he thought he was dreaming and quickly accepted the job. There were many such cases. At that time I was told that three persons "owned" the American Mathematical Society: Oswald Veblen, G. D. Birkhoff, and Arthur B. Coble, from Illinois. Most academic positions were secured through the recommendations of these three. What a contrast to the vast number of university jobs in mathematics which exist today!

It was Veblen who was responsible for Johnny's presence at the Institute. He had invited him for a semester's stay at first, and later arranged for him to remain. He liked Johnny very much and considered him almost as a son.

Oswald Veblen, a nephew of Thorstein Veblen, author of *The Theory of the Leisure Class,* was a famous American mathematician, tall, slim, Scandinavian-looking, with a caustic sense of humor. He was well known for his work on the foundations of projective geometry and topology.

Veblen organized walks in the Princeton woods, and I was invited to join some of these expeditions during which there was lots of mathematical talk and gossip while he cut dead wood and tree branches to clear the paths.

To my mind the Princeton woods with their thin, spindly trees and marshes were not at all impressive compared to the Polish forests. But it was the first time I heard and saw what seemed like gigantic Kafkaesque frogs. The birds were also very different, and I felt truly on another continent, in a very exotic land.

Throughout these walks and discussions there always lurked in the back of my mind a question: would I receive an invitation from some American institution, which would enable me to remain? More subconsciously than consciously, I was eagerly looking for a way to stay, the reason being the critical political situation in Europe and the catastrophic job situation for mathematicians there, especially Jews. There was very little future for me in Poland and it was becoming increasingly evident that the country was in mortal danger. I also admired the freedom of expression, of work, the sense of initiative, the spirit which was in the air; here the future of the world was beckoning. Even though I did not mention any of this explicitly to Johnny, I became very eager to stay and to take work if it presented itself.

About that time, Kuratowski appeared in Princeton for a month's visit. He arrived late in the spring, and wondered whether there was any chance that I might be invited to remain in the States for the next academic year. He went to Harvard to give a lecture, and several of the professors there, Birkhoff, Graustein, and others, asked him about me. He probably gave the best references. When he spoke to me about these possibilities, he had mixed feelings. He knew full well that there was very little chance of a professorship for me in Poland, and realized it was good for my future career to remain in the States a while longer, yet he was genuinely sorry at the thought that I might not return.

During this stay, he and Johnny obtained some very strong results on certain types of projective sets. This is a very elegant theory of operations in mathematical logic, going beyond the Aristotelian or Boolean ones. To this day the theory is full of mysterious situations very fundamental to problems in the foundations of mathematics and set theory. Much recent work concerns such projective operations, and some recent results certainly originate from this interesting paper. It is curious how this came about. With his technical virtuosity and depth of penetration, once Johnny had received the starting impulse he was able to find the decisive points. A good example of how collaboration in mathematics is very often fruitful!

Von Neumann invited me one day to give a talk in his seminar on my results in "semi-simple groups," which was a subject I did not know very much about. I have often succeeded in obtaining rather original and not unimportant results in areas where I did not know the foundations or the details of a theory too well. At this seminar Johnny asked me some very searching and penetrating questions, and I had to think very hard to give satisfactory answers; I did not feel he was doing this to embarrass me, but only because of his overriding objectivity and desire to make things clear.

At some lectures von Neumann sometimes "snowed" the students by elaborating the easier points and quickly glossing over the difficulties, but he always demonstrated his fantastic and to some extent prophetic range of interests in mathematics and its applications and at the same time an objectivity which I admired enormously.

As a mathematician, von Neumann was quick, brilliant, efficient, and enormously broad in scientific interests beyond mathematics itself. He knew his technical abilities; his virtuosity in following complicated reasoning and his insights were supreme; yet he lacked absolute self-confidence. Perhaps he felt that he did not have the power to divine new truths intuitively at the highest levels or the gift for a seemingly irrational perception of proofs or formulation of new theorems. It is very hard for me to understand this. Perhaps it was because on a couple of occasions he had been anticipated, preceded, or even surpassed by others. For instance, he was disappointed that he had not first discovered Gödel's undecidability theorems. He was more than capable of this, had he admitted to himself the possibility that Hilbert was wrong in his program. But it would have meant going against the prevailing thinking of the time. Another example is when G. D. Birkhoff proved the ergodic theorem. His proof was stronger, more interesting, and more self-contained than Johnny's.

During my stay in Princeton I felt that there was some hesitation on Johnny's part about his own work. He was immersed in his new work on continuous geometries and in the theory of classes of operators in Hilbert spaces. I myself was not so interested in problems concerning properties of Hilbert spaces. Johnny, I could feel, was not completely certain of the importance of this work, either. Only when he found, from time to time, some ingenious, technically elegant trick or a new approach did he seem visibly stimulated or relieved of his own internal doubts.

It was at that time that he began to think of problems away from pure mathematics, although this was not the first time in his life. (He had written his famous book on the mathematical foundations of quantum theory in 1929.) He was thinking now more about classical problems in physics. For example, he studied problems of turbulence in hydrodynamics. In his continuous geometries, their elements do not play the role of what we normally consider as "points" in Euclidean space; it is a creation of a "pointless" geometry, a name which lent itself to many an easy joke.

He came back again and again to the possibilities of reformulating the logic of quantum theory, the substance of the talk he gave at the Warsaw seminar. In Princeton he frequently worked on this topic. Listening to his conversation, I saw or felt his own hesitation, and I suffered from doubt, too, because there was no definite experimental possibility of verifying this—it seemed purely a question of logic. Purely "grammatical" approaches never interested me much. When something is merely convenient or typographically useful, it seems less interesting to me than when there is a more real physical base, or if abstract, still somehow palpable. I have to admit that there are cases where formalism by itself has great value—for example, the technique, or rather the notation of Feynman graphs in physics. It is a purely typographical idea, it does not bring in itself any tangible input into a physical picture, nevertheless, by being a good notation it can push thoughts in directions that may prove useful or even novel and decisive. Beyond this (and extremely important), there is the magic of "algorithms," or symbolism in mathematics. Calculus itself shows the wonder of it. Various transforms, generating functions, and the like perform almost miraculously in mathematical applications.

Von Neumann was the master of, but also a little bit the slave to, his own technique. When he saw that something could be done, he let himself be carried away on tangents. My own feeling is that some of his mathematical work on classes of operators or on quasi-periodic functions, for example, is very interesting technically, but to my taste not terribly important; he could not resist doing it because of his facility.

How terribly important habit is. It may largely determine the characteristics or the nature of the brain itself. Habits influence or perhaps can largely determine the choice of trains of thought in one's work. Once these are established (and in my opinion they may be established very quickly—sometimes after just a few trials), the "connections" or "programs" or "subroutines" become fixed. Von Neumann had this habit of considering the line of least resistance. Of course, with his powerful brain he could quickly vanquish all small obstacles or difficulties and then go on. But if the difficulty was great right from the start, he would not knock his head against the wall, nor would he—as I once expressed it to Schreier—walk around the fortress and knock here and there to find the weakest spots and try to break through. He would switch to another problem. On the whole in his work habits I would call Johnny more realistic than optimistic.

Johnny was always a hard worker; he had a great energy and toughness behind a physical appearance that was somewhat on the soft side. Each day he would start writing before breakfast. Even at parties in his house, he would occasionally leave the guests to go to his study for half an hour or so to record something that was on his mind.

He may not have been an easy person to live with—in the sense that he did not devote enough time to ordinary family affairs.

Some people, especially women, found him lacking in curiosity about subjective or personal feelings and perhaps deficient in emotional development. But in his conversations with me, I felt that only a certain shyness prevented him from having more explicit discussions along these lines. Such seeming diffidence is not uncommon among mathematicians. Non-mathematicians often reproach us for this and may resent this apparent emotional insensitivity and excessive quantitative and rational bent, especially in attitudes towards mundane matters outside science. Von Neumann was so busy with mathematics, physics, and with academic affairs, not to mention increasingly innumerable activities later on as a consultant to many projects and Government advisory work, he probably could not be a very attentive, "normal" husband. This might account in part for his not-too-smooth home life.

To be sure, he was interested in women, outwardly, in a peculiar way. He would always look at legs and the figure of a woman. Whenever a skirt passed by he would turn and stare—so much so that it was noticed by everyone. Yet this was absentmindedly mechanical and almost automatic. About women in general he once said to me, "They don't do anything very much." He meant, of course, nothing much of importance outside of their biological and physiological activities.

He did not show social prejudice and never concealed his Jewish origins (even though I think he had actually been baptized a Christian in his childhood). In fact, he was very proud of the birth of the state of Israel in 1948 and was pleased by the Jewish victories over the surrounding Arab countries—a sort of misplaced nationalism.

His father, a banker, had been titled "von." In the Austro-Hungarian Empire people were rewarded with titles but these could also be obtained by gifts of money to the government. Johnny never used the full title (neither did von Kárman who was also of Jewish origin). He was ill at ease with people who were self-made or came from modest backgrounds. He felt most comfortable with third- or fourth-generation wealthy Jews. With someone like me, he would then often use Jewish expressions or jokes as a spice to conversation. He was a man of the world, not exactly snobbish but quite conscious of his position, who felt more at ease with people with the same background.

He was broadly educated and well versed in history, especially of the Roman Empire—its power and organization fascinated him. Perhaps part of this interest stemmed from a mathematician's appreciation of the difference between variables involving individual points, or persons, and groups of such, or classes of things. He was given to finding analogies between political problems of the present and of the past. Sometimes, the analogy was genuinely there, but there were so many other different factors that I don't think his conclusions were always justified.

In general, he tended not to disagree with people. He would not contradict or dissuade when asked for advice about things they were inclined to do. In matters of ordinary human affairs, his tendency was to go along, even to anticipate what people wanted to hear. He also had the innocent little trick of suggesting that things he wanted done originated with the persons he wanted to do them! I started using this ploy myself after I learned it from him. However, in scientific matters, he did defend the principles he believed in.

When it came to other scientists, the person for whom he had a deep admiration was Kurt Gödel. This was mingled with a feeling of disappointment at not having himself thought of "undecidability." For years Gödel was not a professor at Princeton, merely a visiting fellow, I think it was called. Apparently there was someone on the faculty who was against him and managed to prevent his promotion to a professorship. Johnny would say to me, "How can any of us be called professor when Gödel is not?" When I asked him who it was who was unfriendly to Gödel, he would not tell me, even though we were close friends. I admired his discretion.

As for Gödel, he valued Johnny very highly and was much interested in his views. I believe knowing the importance of his own discovery did not prevent Gödel from a gnawing uncertainty that maybe all he had discovered was another paradox à la Burali Forte or Russell. But it is much, much more. It is a revolutionary discovery which changed both the philosophical and the technical aspects of mathematics.

When we talked about Einstein, Johnny would express the usual admiration for his epochal discoveries which had come to him so effortlessly, for the improbable luck of his formulations, and for his four papers on relativity, on the Brownian motion, and on the photo-electric quantum effect. How implausible it is that the velocity of light should be the same emanating from a moving object, whether it is coming toward you or whether it is receding. But his admiration seemed mixed with some reservations, as if he thought, "Well, here he is, so very great," yet knowing his limitations. He was surprised at Einstein's attitude in his debates with Niels Bohr—at his qualms about quantum theory in general. My own feeling has always been that the last word has not been said and that a new "super quantum theory" might reconcile the different premises.

I once asked Johnny whether he thought that Einstein might have developed a sort of contempt for other physicists, including even the best and most famous ones—that he had been deified and lionized too much. No one tried to go him one better by generalizing his theory of relativity, for example, or inventing something which would rival or change or improve it. Johnny agreed. "I think you are right," he said, "he does not think too much of others as possible rivals in the history of physics of our epoch."

Comparisons are invidious, and there is no question of any linear order of eminence or greatness in science. Much of it is a question of taste. It is probably as difficult to compare mathematicians linearly or otherwise as it would be to compare musicians, poets, or writers. There are, of course, large and obvious differences in "class." One could safely say, I think, that Hilbert was probably a greater mathematician than some young teaching assistant chosen at random at a large university. I feel that some of the most permanent, most valuable, most interesting work of von Neumann came towards the end of his life, involving his ideas on computing, on the applications of computing, and on automata. Therefore, when it comes to lasting impact, I think in many ways it might be as great as that of Poincaré's, who was, of course, quite theoretical and did not actually contribute directly to technology itself. Poincaré was one of the great figures in the history of mathematics. So was Hilbert. As mathematicians' mathematicians, they are idolized, perhaps a little more than von Neumann. But final judgments have to be left to the future.

One of the luckiest accidents of my life happened the day G. D. Birkhoff came to tea at von Neumann's house while I was visiting there. He seemed to have heard about me from his son Garrett, whom I had met in Warsaw. We talked and, after some discussion of mathematical problems, he turned to me and said, "There is an organization at Harvard called the Society of Fellows. It has a vacancy. There is about one chance in four that if you were interested and applied you might receive this appointment." Johnny nodded eagerly in my direction, and I said, "Yes, I might be interested in spending some time at Harvard." A month later, in April of 1936, I received an invitation to give a talk at the mathematics colloquium there. The talk was followed by an invitation to a dinner at the Society of Fellows. I suppose this was to look me over without my being aware of it.

At the colloquium, I talked about something which is still being worked on, the existence in many structures of a small number of elements which generate subgroups or subsystems dense in the whole structure. (Or, popularly speaking, out of an infinite variety of objects one can pick a few such that by combining them one can obtain, with only a small error, all the others.) The results were something that Jozef Schreier and I had proved a couple of years earlier. I talked with confidence—I don't remember ever being very nervous about giving talks because I always felt I knew what I was talking about. It must have been well received, for when I returned to Princeton, I found a letter which gladdened me no end. It was from the Secretary of the Harvard Corporation, signed in the English manner "Your Obedient Servant." It was a nomination to the position of Junior Fellow to begin the following autumn and to last for three years. The conditions were extremely attractive: fifteen hundred dollars a year plus free board and room together with some travel allowances. In those days it seemed a royal offer.

With this in my pocket, I happily began preparations to return to Poland for the summer. To make up for Johnny's disaster of the summer before, my new Princeton acquaintances gave me an order to bring back a huge amount of caviar. Little did they realize that in Poland, which did not produce it, it was as expensive as in the West.

CHAPTER 5

Harvard Years

1936–1939

I CAME to the Society of Fellows during its first few years of existence. Garrett Birkhoff and B. F. Skinner, the psychologist, were among its original members. Most of the Junior Fellows, as we were called, were in their mid-twenties, mainly budding post-doctoral scholars.

I was given a two-room suite in Adams House, next door to another new fellow in mathematics by the name of John Oxtoby. About my age, he did not have his doctor's degree but was well known at the University of California—where he had done his graduate work—for his brilliance and promise. I took an instant liking to him. He was a tallish, blue-eyed redhead, with a constant good disposition. An attack of polio in his high-school years had severely crippled one leg, so that he had to walk with a crutch.

He was interested in some of the same mathematics I was: in set theoretical topology, analysis, and real function theory. Right off, we started to discuss problems concerning the idea of "category" of sets. "Category" is a notion in a way parallel to but less quantitative than the measure of sets—that is, length, area, volume, and their generalizations. We quickly established some new results, and the fruits of our conversations during the first few months of our acquaintance were published as two notes in *Fundamenta*. We followed this with an ambitious attack on the problem of the existence of ergodic transformations. The ideas and definitions connected with this had been initiated in the nineteenth century by Boltzmann; five years before work on this had culminated in von Neumann's paper, followed (and in a way superseded) by G. D. Birkhoff's more imposing result. Birkhoff, in his trail-breaking papers and in his book on dynamical systems, had defined the notion of "transitivity." Oxtoby and I worked on the completion to the existence of limits in the ergodic theorem itself.

In order to complete the foundation of the ideas of statistical mechanics connected with the ergodic theorem, it was necessary to prove the existence, and what is more, the prevalence of ergodic transformations. G. D. Birkhoff himself had worked on special cases in dynamical problems, but there were no general results. We wanted to show that on every manifold (a space representing the possible states of a dynamical system)—the kind used in statistical mechanics—such ergodic behavior is the rule.

The nature, intensity and long duration of our daily conversations reminded me of the way work had been done in Poland. Oxtoby and I usually sat in my room, which was rather stark, although I had rented a couple of oriental rugs to furnish it, or in his own, which was even more spartan.

We discussed various approaches to a possible construction of these transformations. With my usual optimism, I was somehow sure of our ultimate success. We kept G. D. Birkhoff informed of the status of our attacks on the problem. He would smile when I talked to him at dinner at the Society of Fellows, partly amused, partly impressed by our single-minded persistence, and partly skeptical, though he really had an open mind about our chances. He would check what I told him with Oxtoby, a more cautious person. It took us more than two years to break through and to finish a long paper, which appeared in *The Annals of Mathematics* in 1941 and which I consider one of the more important results that I had a part in.

The chairman of the Society was L. J. Henderson, a famous biologist, author of a book, *The Fitness of Environment,* which enjoyed a great popularity at the time, not only among specialists, but quite generally. L.J., as he was called, was a great Francophile. Indeed, the Society was molded along the lines of the Fondation Thiers in Paris, rather than on the Cambridge or Oxford systems of fellows in the college.

The Society was composed of some five or six Senior Fellows and about twenty-two Junior Fellows.

The Senior Fellows were well-known distinguished professors, like John Livingston Lowes in literature, Samuel Eliot Morison, the historian, Henderson, and Alfred North Whitehead, the famous English philosopher, who had already retired from his professorship at Harvard when I entered the Society. I often had the pleasure of sitting next to him at the traditional Monday-night dinners of the Society.

Some of the Junior Fellows gave me the impression of being a somewhat precious group of young men, as far as manners were concerned. Oxtoby, Willard Quine (really a logician), and I were the only mathematicians among them. Among the physicists there were several who later became very well known, such as John Bardeen, Ivan Getting, and Jim Fisk. Among the biologists, I remember Robert B. Woodward, the chemist who first synthesized quinine and other important biological substances. Paul Samuelson, the economist who served as advisor to President Kennedy, was there; also Ivar Einerson, a great scholar in linguistics; Henry Guerlach, who became a historian of science; and Harry Levin, in English literature. Levin was rather proustian in his manner. He loved to engage in sophisticated and what seemed to me occasionally rather precious discussions. Another foreign-born member was George Hanfmann, an archaeologist. Hanfmann was obviously a very learned person, and I appreciated his erudition. We shared the same fondness for Greek and Latin literature.

The logician Willard Quine was friendly and outgoing. He was interested in foreign countries, their culture and history, and knew a few words of Slavic languages, which he used on me with great gusto. He already had made a reputation in mathematical logic. I remember him as slim, dark-haired, dark-eyed—an intense person. During the presidential election of 1936 in which Franklin D. Roosevelt defeated Landon, I met him on the stairs of Widener Library at nine in the morning, after Roosevelt's landslide victory. We stopped to chat and I asked him: "Well, what do you think of the results?" "What results?" he replied. "The presidential election, of course," I said. "Who is President now?" he asked casually. This was characteristic of many in academe. I once heard that, during Charles W. Eliot's presidency at Harvard, a visitor to his house was told, "The President is away in Washington to see Mr. Roosevelt"! (This was Theodore Roosevelt.)

I had my meals at Adams House, and the lunches there were particularly agreeable. We sat at a long table—young men and sometimes great professors; the conversations were very pleasant. But often, towards the end of a meal, one after the other would gulp his coffee and suddenly announce: "Excuse me, I've got to go to work!" Young as I was I could not understand why people wanted to show themselves to be such hard workers. I was surprised at this lack of self-assurance, even on the part of some famous scholars. Later I learned about the Puritan belief in hard work—or at least in appearing to be doing hard work. Students had to show that they were conscientious; the older professors did the same. This lack of self-confidence was strange to me, although it was less objectionable than the European arrogance. In Poland, people would also pretend and fabricate stories, but in the opposite sense. They might have been working frantically all night, but they pretended they never worked at all. This respect for work appeared to me as part of the Puritan emphasis on action versus thought, so different from the aristocratic traditions of Cambridge, England, for example.

The Society's rooms were in Eliot House. We Junior Fellows would meet there on Mondays and Fridays for lunch, and for the famous Monday-night dinners which gathered Junior and Senior Fellows together around a long T-shaped table which was said to be the one featured in Oliver Wendell Holmes' *Autocrat of the Breakfast-Table.* Henderson had secured it from some Harvard storeroom.

President Lowell attended almost every Monday dinner. He was fond of re-creating the Battle of Jutland of World War I, moving knives and forks and saltcellars around on the dinner table to show the positions of the British and German fleets. From time to time he would also betray his doubts and even remorse about the Sacco and Vanzetti case. He would recount it—not so much to defend but rather to restate the position of the court and the subsequent legal steps. He had been a member of one of the review committees.

Good French Burgundies or Alsatian wines accompanied the meals. These were the pride and joy of Henderson, who once told me that if he ever deserved a statue in Cambridge, he would like to be put in Harvard Square with a bottle of wine in his hands, in commemoration of his having been the first person to obtain University funds for a wine cellar. George Homans, one of the Junior Fellows, a descendant of President John Adams, was one of the young men entrusted with the selection and sampling of wines. I considered it a great distinction when I, too, was put on the wine-tasting committee of the Society. This was my very first administrative job in America! The Society is still very much alive today at Harvard, and it continues to hold its Monday-night dinners where former fellows are always welcome.

In 1936 the depression appeared to be ending. Harvard

University seemed relatively untouched by this cataclysm. After the colloquium talk I gave there just before my appointment to the Society of Fellows, I remember Professor William Graustein telling me that at Harvard the professors had not felt the depression at all. This left me wondering at their lack of involvement in the general problems of the country or in the affairs of Massachusetts or even of Cambridge. It was evident that campus life in America meant at least partial isolation from the rest of society. Professors lived almost entirely among themselves and had very little contact with the rest of the professional or creative community as in Lwów. This had both good and bad effects: more time for scholarly work, but very little influence on the life of the country or vice versa. As everyone knows, things changed somewhat after World War II. In the Kennedy administration, for example, Harvardians had a great deal to do with the affairs of government and for a time the influence of scientists became even paramount.

Activities at the Society of Fellows were of course only one facet of my life at Harvard. I had many contacts with the younger members of the faculty at the university and quite often saw and talked with the senior professors and with G. D. Birkhoff himself. His son Garrett, a tall, good-looking, and brilliant mathematician, some two years younger than I, became a friend, and we saw each other nearly every day.

Even though membership in the Society did not require teaching of any kind, Professor Graustein asked me to teach an elementary undergraduate section of a freshman course called Math 1A. (It may even be that the late President Kennedy was for a while a student in this class. I remember a name like that and someone saying that the young man was a rather remarkable person. He left to go abroad in the middle of the term. Years later when I met President Kennedy I forgot to ask him whether he had really taken that course.)

I had given talks and seminars, but not yet taught a regular class, and I found this teaching interesting. The rule for young instructors was to follow very closely the prescribed textbook. Apparently I did not do too badly, for in an evaluation of teachers the student newspaper praised me as an interesting instructor. Soon after the beginning of the course, G. D. Birkhoff came to inspect my performance. Perhaps he wanted to check my English. He sat at the back of the room and watched as I explained to the students how to write equations of parallel lines in analytic geometry. Then I said that next we would study the formulae for perpendicular lines, which, I added, were "more difficult." After the lecture Birkhoff came to me and commented, "You've done very well, but I would not have said that perpendicular lines are more difficult." I replied that I believed on the contrary that students would remember better this way than if I said everything was easy. Birkhoff smiled at this attempt at pedagogy on my part. I think he liked my independence and outspoken ways, and I saw him rather frequently.

Shortly after I arrived in Cambridge, he had invited me to dine at his house. It was my first introduction to strange dishes like pumpkin pie. After dinner, which was pleasant enough, I got ready to leave and G.D. took my overcoat to help me into it. This sort of courtesy was unheard of in Poland; an older man would never have helped a much younger one. I remember blushing crimson with embarrassment.

I frequently ate lunch with his son Garrett, and we often took walks together. We talked much about mathematics and also indulged in the usual gossip that mathematicians love. Surely it is a shallow theme to evaluate how good X or Y is, but it is a characteristic of our tribe. The reader may have noticed that I practice this, too. Mathematics being more in the nature of an art, values depend on personal tastes and feelings rather than on objective factual notions. Mathematicians tend to be rather vain—though less so than opera tenors or artists. But as every mathematician knows some special bit of math better than anyone else, and math is such a vast and now more and more specialized subject, some like to propose linear orders of "class" among the better-known ones and to comment on their relative merits. On the whole, it is a harmless if somewhat futile pastime.

I remember that at the age of eight or nine I tried to rate the fruits I liked in order of "goodness." I tried to say that a pear was better than an apple, which was better than a plum, which was better than an orange, until I discovered to my consternation that the relation was not transitive— namely, plums could be better than nuts which were better than apples, but apples were better than plums. I had fallen into a vicious circle, and this perplexed me at that age. Mathematicians' ratings are something like this.

Many mathematicians are also sensitive about what they consider their most beautiful mental offspring—results or theorems—and they tend to be possessive about them. Para- doxically, they also show a tendency to consider their own work as difficult and other work as easier. This is exactly op- posite in other fields where the better acquainted one is with something the easier it seems.

Mathematicians are also prone to disputes, and per- sonal animosities between them are not unknown. Many years later, when I became chairman of the mathematics department at the University of Colorado, I noticed that the difficulties of administering N people was not really propor- tional to N but to N^2. This became my first "administrative theorem." With sixty professors there are roughly eighteen hundred pairs of professors. Out of that many pairs it was not surprising that there were some whose members did not like one another.

Among the Harvard mathematicians I knew, I should mention Hassler Whitney, Marshall Stone, and Norbert Wiener. Whitney was a young assistant professor, interest- ing not only as a mathematician. He was friendly, but rather taciturn—psychologically of a type one encounters in this country more frequently than in central Europe—with wry humor, shyness but self-assurance, a probity which shines through, and a certain genius for persistent and deep follow- through in mathematics.

Marshall Stone, whom I had met when he came through Warsaw with von Neumann and Birkhoff in 1935 on the way back from the Moscow Congress, had had a meteoric career at the university, although he was only thirty-one years old. Already a full professor, he was quite influential in the affairs of the department and of the university for that matter. He wrote a classic work, a comprehensive and authoritative book on Hilbert space, an infinitely dimensional generalization of the three-dimensional or n-dimensional Euclidean space, mathematically basic to modern quantum theory in physics. He was the son of Harlan Stone, Chief Justice of the Supreme Court. It is said that his father proudly said of Marshall's mathematical achievements, "I am puzzled but happy that my son has written a book of which I understand nothing at all."

And there was Norbert Wiener! I met him at a colloquium talk I gave during my first year at Harvard. I was lecturing on some problems of topological groups, and mentioned a result I had obtained in Poland in 1930 on the impossibility of completely additive measure defined in all subsets of a given set. Wiener, who always sat at lectures in a semi-somnolent state except when he heard his name (at which he would suddenly jump up, then sit back in a very comical way) interrupted me to say, "Oh! Vitali has proved something like that already." I replied that I knew Vitali's result and that it was much weaker than mine because it required an additional property—namely equality for congruent sets—whereas my result did not make any such postulate and was a much stronger, purely set theoretical proof. After the lecture he came to me, apologized, and agreed with my statements. This was the beginning of our acquaintance.

I had heard of Wiener before this meeting, of course, not only about his mathematical wizardry, his work in number theory, his famous Tauberian theorems, and his work on Fourier Series, but also about his eccentricities. In Poland, I had heard through Jozef Marcinkiewicz about his book with

Paley on the summability of Fourier transforms. Raymond Paley, one of the most promising and successful young English mathematicians, died in a mountaineering accident at a tragically young age. Marcinkiewicz was a student of Antoni Zygmund. He visited Lwów as a post-doctoral fellow and patronized the Scottish Café, where we discussed Wiener's work, since he had worked in trigonometric series, trigonometric transforms, and summability problems. Marcinkiewicz, like Paley, whom he resembled in genius and in mathematical interests and accomplishments, reportedly was killed while an officer in the Polish army in the 1939 campaign at the beginning of World War II.

Absentminded and otherworldly in appearance, Wiener nevertheless could make an intuitive appraisal of others, and he must have been interested in me. Great as the difference in age between us was (his forty to my twenty-six years), he would seek me out occasionally in my little apartment in Adams House, sometimes late in the evening, and propose a mathematical conversation. He would say, "Let's go to my office, where I can write on the blackboard." This suited me better than staying in my rooms, from which it would have been difficult to put him out without being rude. So he drove me in his car through darkened streets to MIT, opened the building doors, turned on the light, and he started talking. After an hour or so, although Wiener was always interesting, I would almost fall asleep and finally manage to suggest that it was time to go home.

Wiener seemed childish in many ways. Being very ambitious about his place in the history of mathematics, he needed constant reassurance about his creative ability. I was almost stunned a few weeks after our first encounter when he asked me point blank: "Ulam! Do you think I am through in mathematics?" Mathematicians tend to worry about their diminishing power of concentration much as some men do about their sexual potency. Impudently, I felt a strong temptation to say "yes" as a joke, but refrained; he would not have understood. Speaking of that remark, "Am I through," several years later at the first World Congress of Mathematicians held in Cambridge, I was walking on Massachusetts Avenue and saw Wiener in front of a bookstore. His face was glued to the window and when he saw me, he said, "Oh! Ulam! Look! There is my book!" Then he added, "Ulam, the work we two have done in probability theory has not been noticed much before, but see! Now, it is in the center of everything." I found this disarmingly and blessedly naive.

Anecdotes about Wiener abound; every mathematician who knew him has his own collection. I will add my story of what happened when I came to MIT as a visiting professor in the fall of 1957. I was assigned an office across the hall from his. On the second day after my arrival, I met him in the corridor and he stopped me to say, "Ulam! I can't tell you what I am working on now, you are in a position to put a secret stamp on it!" (This presumably because of my position in Los Alamos.) Needless to say, I could do no such thing.

Wiener always had a feeling of insecurity. Before the war he used to talk about his personal problems to J. D. Tamarkin, who was a great friend of his. When he was writing his autobiography, he showed a voluminous manuscript to Tamarkin. Tamarkin, whom I had met in 1936 and with whom I became quite friendly, told me about Wiener's manuscript and how interesting it was. But he also expressed the opinion that Wiener might be sued for libel for many of his outspoken statements. He spoke almost with disbelief about Wiener's text and how he tried to dissuade him from publishing the book in that form. What finally appeared apparently was considerably toned down from the original version.

Another memory I have of Wiener concerns his asking me to go with him to South Station in Boston to meet the English mathematician G. H. Hardy who was coming to the States for a visit. He knew I had met Hardy in England. We collected another mathematician, perhaps it was Norman Levinson, and picked up Hardy at the train. Wiener, who prided himself on his knowledge of the Chinese, their culture and even their language, invited everybody for lunch at a Chinese restaurant. Immediately he started talking Chinese to the waiter, who seemed not to understand a word. Wiener simply remarked, "He must be from the south and does not speak Mandarin." (We were not quite convinced that this was the complete explanation.) It was a very pleasant lunch with much mathematical talk. And after lunch Wiener who had picked up the check discovered that he had no money. Fortunately we found the few necessary dollars in our pockets. Wiener scrupulously reimbursed us later.

It was said that Wiener, although he considered his professorship at MIT quite satisfactory, was very disappointed that Harvard never offered him a post. His father had been a professor at Harvard, and Norbert wanted very much to follow in his footsteps.

Although G. D. Birkhoff was at least ten years his senior, Wiener felt a rivalry with him and wanted to equal or surpass him in mathematical achievement and fame. When Birkhoff's celebrated ergodic theorem proof was published, Wiener tried very hard to go him one better and prove an even stronger theorem. He did manage it, but the strengthening was not as simple or as fundamental as G.D.'s original proof. Here again is an example of the competitive nature of some mathematicians and the sources of their ambition.

I think Wiener had marvelous talents as a mathematician—that is perspicacity and technical genius. He had a supreme general intelligence but, in my opinion, not the spark of originality which does the unusual unrelated to what others have done. In mathematics, as in physics, so much depends on chance, on a propitious moment. Perhaps von Neumann also lacked some of the "irrational," though with his wonderful creativity, he certainly went to and achieved the limits of the "reasonable."

There are several ways in which Wiener and von Neumann intersected in their interests and in their feelings about what was important both in pure mathematics and its applications, but it is difficult to compare their personalities. Norbert Wiener was a true eccentric and von Neumann was, if anything, the opposite—a really solid person. Wiener had a sense of what is worth thinking about, and he understood the possibilities of using mathematics for seemingly more important and more visible applications in theoretical physics. He had a marvelous technique for using Fourier transforms, and it is amazing how much the power of algorithms or symbolism could accomplish. I am always amazed how much a certain facility with a special and apparently narrow technique can accomplish. Wiener was a master at this. I have seen other mathematicians who could do the same in a more modest way. For instance, Steinhaus obtained quite penetrating insights into other fields, and his student, Mark Kac, now at Rockefeller University, surpassed him. Antoni Zygmund in Chicago, another Pole, is a master of the great field of trigonometric series. Several of his students have obtained epoch-making results in other fields—for example, Paul Cohen, who did this in set theory, the most general and abstract part of mathematics.

I don't think Wiener was particularly fond of combinatorial thinking or of working on foundations of mathematico-logical or set theoretical problems. At the beginning of his career, he may have gone in this direction, but later he applied himself to other fields and to number theory.

Von Neumann was different. He also had several quite independent techniques at his fingertips. (It is rare to have more than two or three.) These included a facility for symbolic manipulation of linear operators. He also had an undefinable "common sense" feeling for logical structure and for both the skeleton and the combinatorial superstructure in new mathematical theories. This stood him in good stead much later, when he became interested in the notion of a possible theory of automata, and when he undertook both the conception and the construction of electronic computing machines. He attempted to define and to pursue some of the formal analogies between the workings of the nervous system in general and of the human brain itself, and the operation of the newly developed electronic computers.

Wiener, somewhat hemmed in by the childishness and naiveté of his personality, was perhaps psychologically handicapped by the fact that, as a child, his father had pushed him as a prodigy. Von Neumann, who also began rather young, had a much wider knowledge of the world and more common sense outside the realm of pure intellect. Furthermore, Wiener was perhaps more in the tradition of talmudistic Judaic scholarship, even though his opinions and beliefs were very libertarian. This was quite conspicuously absent from von Neumann's makeup.

Johnny's overwhelming curiosity included many fields of theoretical physics, beginning with his pioneering work—his attempt to form a rigorous mathematical basis for quantum theory. His book, *Die Mathematische Grundlagen der Quantum Mechanik,* published over forty years ago, is not only a classic, but still the "bible" on the subject. He was especially fascinated by the puzzling role of the Reynolds number and the seeming mystery of sudden onsets of turbulence in the motions of fluids. He had discussions with Wiener on the perplexing values of this number which is "dimensionless"—a pure number expressing the ratio of the inertial forces to the viscous forces. It is of the order of two thousand, a large number. Why is this so and not around one, or ten, or fifty? At that time, Johnny and I came to the conclusion that actual detailed numerical computations of many special cases could help throw light on the reasons for the transition from a laminar (regular) to a turbulent flow.

He told me of another discussion he had with Wiener and their different points of view: Johnny advocated, in order to establish models for the working of the human brain, a numerical digital approach through a sequence of time steps, while Wiener imagined continuous or "hormonal" outlines. The dichotomy between these points of view is still of great interest and, of course, by now has been transformed and deepened by the greater knowledge of the anatomy of the brain and by more work in the theory of automata.

The relationship between G. D. Birkhoff and von Neumann was curious. Birkhoff did not really have complete admiration for or appreciation of von Neumann's genius. He probably could not appreciate the many kinds of mathematics von Neumann was pursuing. He admired his technical brilliance, but G.D.'s tastes were more classical, in the tradition of Poincaré and the great French school of analysis. Von Neumann's interests were different. Birkhoff had ambitions to produce something of great importance in physics, and he made a few technically interesting but not conceptually important contributions to the general theory of relativity. He lectured several times on such subjects in Mexico, stimulating a small school of relativists there. Von Neumann's interests lay in the foundations of the new quantum theory's more recent developments. Theirs were differences of interests, of approaches, and of value systems. Birkhoff appreciated probing in depth more than exploring in breadth. Von Neumann, to some extent, did both. There was, of course, about a quarter-century's difference in age between them, as well as in background and in upbringing. Also, von Neumann never quite forgave G.D. for having "scooped" him in the affair of the ergodic theorem: Von Neumann had been first in proving what is now called the weak ergodic theorem. By a sheer virtuoso kind of combinatorial thinking, Birkhoff managed to prove a stronger one, and—having more influence with the editors of the *Proceedings* of the National Academy of Sciences—he published his paper first. This was something Johnny could never forget. He sometimes complained about this to me, but always in a most indirect and oblique way.

In addition to the elementary mathematics courses which I taught during my first year in the Society, I was asked to add advanced courses gradually. I liked this, for the best way to learn a subject is to try to teach it systematically.

Then one gets the real points, the essentials. One was an important undergraduate course in classical mechanics, Math 4 if I remember its former name. Another was Math 9, a course on probability.

At the time I had no precise idea what grades meant: A, B, C, D, or F. But I had rigid standards. I remember an otherwise quite good student, who protested receiving the grade of "C." Some other professors intervened, but I stubbornly, perhaps foolishly, stood my ground. Now I tend to be more lenient, and when I give a "C" or "D" the students really deserve an "F" or worse!

Tamarkin, who was a professor at Brown University, asked me to teach a graduate course in his place while he took his sabbatical leave for a term. I decided to give the course on the theory of functions of several real variables. It included a lot of new material—much of it my own recent work—and I was rather proud of it. Every Friday I went to Providence by train, taught the course, spent the weekend with Tamarkin at his home, returning to Cambridge on Sunday. When I mentioned the contents of the course to Mazur when I went home to Lwów for the last time during the summer of 1939, he liked it very much. He liked the material, the way it was organized, and said he would love to give such a course himself, all of which pleased me and encouraged me.

Tamarkin was a most interesting person. He was of medium height, very portly—I would say some thirty pounds overweight. He was quite nearsighted, a constant cigar-and-cigarette-smoker, and generally extremely jovial. As I got to know him better, I discovered the wonderful qualities of his mind and character.

Before World War I, he had written some mathematical research papers on the work of G. D. Birkhoff and even improved some of the latter's results a bit, which led to a certain animosity in their relations. Yet when he came to the United States, Birkhoff helped him secure his position at

Brown, which had a notable mathematics faculty, including James Richardson, Raymond C. Archibald, and others. Richardson was a gentleman of the old school. Archibald was an eminent historian of mathematics, who established the famous mathematical library of Brown, one of the best in the country.

Tamarkin was interested in Polish-style mathematics and had heard about some of my results in the theory of Banach spaces. He had a quality which perhaps only a small number of mathematicians possess: he was extremely interested in the works of others and less egocentric than most. He was also interested in what was going on in other fields besides his own, whereas most mathematicians—even the best ones—are often deeply immersed in their own work and do not pay much attention to what others around them are doing. Tamarkin befriended me and encouraged me in my work.

He was Russian, not of Jewish origin exactly, but a Karaite. The Karaites were a sect of Semitic people not subject to the usual restrictions on Jews in Russia, the reason being that they claimed they were absent from Palestine when Jesus was condemned to death, and this exempted them. This claim was accepted by the Russian governors. They also had something in common with the ancient Khazars, people of a mysterious sixth- or seventh-century kingdom in southern Russia, a pagan tribe whose king decided to adopt a new religion. He selected Judaism after having asked Christian, Moslem, and Jewish representatives to explain their beliefs. Tamarkin believed he was one of their descendants. He had escaped from Leningrad after the Russian Revolution in a manner not unlike that of George Gamov some ten years later—over the ice of Lake Ladoga to Finland.

While I was at Harvard, Johnny came to see me a few times, and I invited him to dinner at the Society of Fellows. We would also take automobile drives and trips together during which we discussed everything from mathematics to literature and talked without interruption while still paying attention to our surroundings. Johnny liked this kind of travel very much.

Once at Christmas time in 1937, we drove from Princeton to Duke University to a meeting of the American Mathematical Society. On the way, among other things we discussed the effect that the arrival of increasingly large numbers of refugee European scientists would have on the American academic scene. We stopped at an inn where we found a folder describing a local Indian Chief, Tomo-Chee-Chee, who apparently had been unhappy about the arrival of white men. As an illustration of our frequently linguistic and philological jokes, I asked him why it was that the Pilgrims had "landed" while the present European immigrants and scientific refugees merely "arrived." Johnny enjoyed the implied contrast and used this in other contexts as an example of an implied value judgment. We also likened G. D. Birkhoff's increasing qualms about the foreign influence to the Indian Chief's. Continuing our drive, we managed to lose our way a couple of times and joked that it was Chief Tomo-Chee-Chee who had magically assumed the shape of false road signs to lead us astray.

This was the first time I visited the South, and I was much taken by the difference in atmosphere between New York, New England, and the southern states. I remember a feeling of "déjà vu": the more polished manners, the more leisurely pace of life, and the elegant estates. Something seemed familiar, and I wondered what it was. Suddenly I asked myself if it could be the remnants of the practice of slavery, which reminded me of the traces of feudalism still visible in the country life of Poland. I was also surprised to see so many black people, and their language intrigued me. At a gas station, one of the Negro attendants said, "What would you like now, Captain?" I asked Johnny, "Does he think I might be an officer and calls me Captain as a compliment?" Similarly, the first time I heard myself called "Doc," I wondered how the porter knew that I had a Doctor's degree!

As we passed the battlefields of the Civil War, Johnny recounted the smallest details of the battles. His knowledge of history was really encyclopedic, but what he liked and knew best was ancient history. He was a great admirer of the concise and wonderful way the Greek historians wrote. His knowledge of Greek enabled him to read Thucydides, Herodotus, and others in the original; his knowledge of Latin was even better.

The story of the Athenian expedition to the island of Melos, the atrocities and killings that followed, and the lengthy debates between the opposing parties fascinated him for reasons which I never quite understood. He seemed to take a perverse pleasure in the brutality of a civilized people like the ancient Greeks. For him, I think it threw a certain not-too-complimentary light on human nature in general. Perhaps he thought it illustrated the fact that once embarked on a certain course, it is fated that ambition and pride will prevent a people from swerving from a chosen course and inexorably it may lead to awful ends, as in the Greek tragedies. Needless to say this prophetically anticipated the vaster and more terrible madness of the Nazis. Johnny was very much aware of the worsening political situation. In a Pythian manner, he foresaw the coming catastrophe.

It was during this trip also that for the first time I sensed that he was having problems at home. He exhibited a certain restlessness and nervousness and would frequently stop to telephone to Princeton. Once he came back to the car very pale and obviously unhappy. I learned later that he had just found out that his marriage to Marietta was definitely breaking up. She would leave him shortly thereafter to marry a younger physicist, one of the frequent guests at the numerous parties which the von Neumanns gave in Princeton.

On the way back from the meeting I posed a mathematical problem about the relation between the topology and the purely algebraic properties of a structure like an abstract group: when is it possible to introduce in an abstract group a topology such that the group will become a continuous topological group and be separable? "Separable" means that there exists a countable number of elements dense in the whole group. (Namely, every element of the group can be approximated by elements of this countable set.) The group, of course, has to be of power continuum at most—obviously a necessary condition. It was one of the first questions which concern the relation between purely algebraic and purely geometric or topological notions, to see how they can influence or determine each other.

We both thought about ways to do it. Suddenly, while we were in a motel I found a combinatorial trick showing that it could not be done. It was, if I say so myself, rather ingenious. I explained it to Johnny. As we drove Johnny later simplified this proof in the sense that he found an example of a continuum group which is even Abelian (commutative) and yet unable to assume a separable topology. In other words, there exist abstract groups of power continuum in which there is no possible continuous separable topology. What is more, there exist such groups that are Abelian. Johnny, who liked verbal games and to play on words, asked me what to call such a group. I said, "nonseparabilizable." It is a difficult word to pronounce; on and off during the car ride we played at repeating it.

Mathematicians have their own brand of "in" humor like this. Generally speaking, they are amused by stories involving triviality of identity of two definitions or "tautologies." They also like jokes involving vacuous sets. If you say something which is true "in vacuo," that is to say, the conditions of the statement are never satisfied, it will strike them as humorous. They appreciate a certain type of logical non sequitur or logical puzzle. For instance, the story of the Jewish mother who gives a present of two ties to her son-in-law.

The next time she sees him, he is wearing one of them, and she asks, "You don't like the other one?"

Some of von Neumann's remarks could be devastating, even though the sarcasm was of an abstract nature. Ed Condon told me in Boulder of a time he was sitting next to Johnny at a physics lecture in Princeton. The lecturer produced a slide with many experimental points and, although they were badly scattered, he showed how they lay on a curve. According to Condon, von Neumann murmured, "At least they lie on a plane."

Some people exhibit an ability to recall stories and tell them to others on appropriate occasions. Others have the ability to invent them by recognizing analogies of situations or ideas. A third group has the ability to laugh and enjoy other people's jokes. I sometimes wonder if types of humor could be classified according to personality. My friends and collaborators, C. J. Everett in the United States and Stanislaw Mazur in Poland, each had a wry sense of humor, and physically and in their handwriting they also resemble each other.

Generally von Neumann preferred to tell stories he had heard; I like to invent them. "I have some wit; it is a tremendous quality," my wife says I once told her. When she pointed out that I was bragging, I promptly added, "True. My faults are infinite, but modesty prevents me from mentioning them all."

In addition to "in" jokes, mathematicians also practice a form of "in" language. For example, they use the word "trivial." It is an expression they are very fond of, but what does it really mean? Easy? Simple? Banal? A colleague of my friend Gian-Carlo Rota once told him that he did not like teaching calculus because it was so trivial. Yet, is it? Simple as it is, calculus is one of the great creations of the human mind, with beginnings dating back to Archimedes. It was "invented" by Newton and Leibnitz, and amplified by Euler, Lagrange, and others. It has a beauty and an importance going far beyond most of the mathematics of our present culture. So what is "trivial"? Certainly not Cantor's great set theory, technically very simple, but deep and wonderful conceptually without being difficult or complicated.

I have heard mathematicians sneer at the special theory of relativity, calling it nothing but a technically trivial quadratic equation and a few consequences. Yet it is one of the monuments of human thought. So what is "trivial"? Simple arithmetic? It may be trivial to us, but is it to the third-grade child?

Let us consider some other words mathematicians use: what about the adjective "continuous"? Out of this one word came all of topology. Topology may be considered as a big essay on the word "continuous" in all its ramifications, generalizations, and applications. Try to define logically or combinatorially an adverb like "even" or "nevertheless." Or take an ordinary word like "key," a simple object. Yet it is an object far from easy to define quasi-mathematically. "Billowing" is a motion of smoke, for example, in which puffs are emitted from puffs. It is almost as common in nature as wave motion. Such a word may give rise to a whole theory of transformations and hydrodynamics. I once tried to write an essay on the mathematics of three-dimensional space that would imitate it.

Were I thirty years younger I might try to write a mathematical dictionary about the origins of mathematical expressions and concepts from commonly used words, imitating the manner of Voltaire's *Dictionnaire Philosophique*.

CHAPTER 6

Transition and Crisis

1936–1940

EACH summer between 1936 and 1939, I returned to Poland for a full three months. The first time, after only a few months' stay in America, I was surprised that street cars ran, electricity and telephones worked. I had become imbued with the idea of America's absolute technological superiority and unique "know-how." My main emotional reactions were, of course, related to reunion with my family and friends, and the familiar scenes of Lwów, followed by a longing to return to the free and hopeful "open-ended" conditions of life in America. To simplify a description of these complicated feelings: in May I started counting the days and weeks left before returning to Europe, then after a few weeks in Poland I would count the days impatiently before I would return to America.

Most mathematicians remained in Lwów during the

summer, and our sessions in the coffee houses and my own personal contacts with them continued until the outbreak of World War II. As before, I worked with Banach and Mazur. Twice, while Banach was spending a few days in Skole or in nearby villages in the Carpathian mountains, some seventy miles south of Lwów, I visited him. I knew these places from my childhood. Banach was working on some of his textbooks, but there was always lots of time to sit in a country inn and discuss mathematics and "the rest of the universe," an expression which was dear to von Neumann. The last time I saw Banach was in late July of 1939 at the Scottish Café. We discussed the likelihood of war with Germany and inscribed a few more problems in the Scottish Book.

In the summer of 1937, Banach and Steinhaus asked me to invite von Neumann to come and give a lecture in Lwów. He arrived from Budapest and spent several days among us. He gave a nice lecture, and I brought him to the Café several times. He jotted some problems in the Scottish Book, and we had some very pleasant discussions with Banach and several others.

I told Banach about an expression Johnny had once used in conversation with me in Princeton before stating some non-Jewish mathematician's result, "Die Goim haben den folgenden Satz bewiesen" (The goys have proved the following theorem). Banach, who was pure goy, thought it was one of the funniest sayings he had ever heard. He was enchanted by its implication that if the goys could do it, Johnny and I ought to be able to do it better. Johnny did not invent this joke, but he liked it and we started using it.

I showed Johnny the city. I had experience in showing it to foreign mathematicians; when I was only a freshman, because I could speak English, Kuratowski had assigned to me the job of showing the town to the American topologist Ayres. I also had to escort Edward Czech, G. T. Whyburn, and several others around Lwów during their Polish visits.

Johnny was much interested in Lwów and surprised at the nineteenth-century appearance of the center of town and its many relics of the fifteenth, sixteenth and seventeenth centuries. Both Hungary and Poland were still semi-feudal in some respects. There were many picturesque parts of town, where old houses leaned towards each other, and crooked narrow cobbled streets. In one little street in the Ghetto, black-market operations in currency were conducted openly. Butcher shops in the suburbs had sides of beef hanging exposed to full view. There were still horse-drawn carriages and electric street-car lines. Taxis were not numerous, and even in the late nineteen-thirties one could take a horse-drawn "fiacre," usually pulled by two horses. When I first arrived in New York, I was surprised to see the shabby old fiacres in front of the best Fifth Avenue hotels with only one poor horse to pull them.

We visited an Armenian church with frescoes by Jan Henryk Rosen, a contemporary Polish artist now in the United States. We also went into a little Russian Orthodox church, where we were both shocked by the sight of a corpse in a half-open coffin about to be buried according to Russian ritual. It was the first time I had ever seen a dead person.

Johnny also came to our house. He met my parents—my mother, who was to die the following year, and my father, who had heard so much about him from me. I took him to see my father's offices, which were in a different part of our big house on Kosciusko Street.

Johnny already knew some of my family. An aunt of mine, the widow of my father's brother Michael, had married a Hungarian financier by the name of Arpad Plesch. Von Neumann knew the Plesches. Arpad's brother Janos was Einstein's physician in Berlin. Arpad was an immensely rich financier, but a rather controversial figure. My aunt was wealthy too, a remarkable woman, descended from a famous fifteenth-century Prague scholar named Caro. In Israel many years later, while I was visiting the town of Safed with von

Kárman, an old Orthodox Jewish guide with earlocks showed me the tomb of Caro in an old graveyard. When I told him that I was related to a Caro, he fell on his knees— that cost me a triple tip. The Plesches traveled frequently and lived often in Paris. I visited them there on my trip in 1934. My aunt's first husband, Michael Ulam, my uncle, was buried in Monte Carlo, and my aunt, who is now dead too, is also buried there in a fantastic marble mausoleum in the Catholic cemetery. Aunt Caro was directly related to the famous Rabbi Loew of sixteenth-century Prague, who, the legend says, made the Golem—the earthen giant who was protector of the Jews. (Once, when I mentioned this connection with the Golem to Norbert Wiener, he said, alluding to my involvement with Los Alamos and with the H-bomb, "It is still in the family!") So much for the family's rich connections.

A story which Johnny and I liked to tell each other, though I do not remember who proposed it first, was a quote from one of those rich uncles, who used to say: "Reich sein ist nicht genug, man musst auch Geld in der Schweiz haben!" (It is not enough to be rich, one must also have money in Switzerland.)

In the summer of 1938, it was von Neumann's turn to invite me to Budapest. I traveled by train via Cracow, and went directly to his house. (I think the address was 16 Arany János Street.) He had reserved a room for me at the Hotel Hungaria, the best hotel in town at that time. It was at the end of a narrow little street, so narrow in fact, that there was a revolving platform at the end for cars to turn around on, as for locomotives in a roundhouse.

Johnny showed me Budapest. It was a beautiful city with the houses of Parliament and the bridges on the river. After dinner at his house, where I met his parents, we went to nightclubs and discussed mathematics! Johnny was alone that year, his marriage was breaking up and Marietta had remained in America.

The next day sitting in a "Konditorei," talking, joking,

and eating, we saw an elegantly dressed lady going by. Johnny recognized her. She entered, and they exchanged a few words. After she left, he explained that she was an old friend, recently divorced. I asked him, "Why don't you marry the divorcée?" Perhaps this implanted the thought in his mind. The next year they indeed were married. Her name was Klara Dan. We became very good friends later on. Johnny and Klari, as she was known to her friends, were married in Budapest and she moved to Princeton in the late summer or fall of 1939. Klari was a moody person, extremely intelligent, very nervous, and I often had the feeling that she felt that people paid attention to her mostly because she was the wife of the famous von Neumann. This was not really the case, for she was a very interesting person in her own right. Nevertheless, she had these apprehensions, which made her even more nervous. She had been married twice before (and married a fourth time after von Neumann's death). She died in 1963 in tragic and mysterious circumstances. After leaving a party given in honor of Nobel Prize-winner Maria Mayer, she was found drowned on the beach at La Jolla, California.

Johnny took me also to a pleasant mountain resort called Lillafüred to visit his former professors, Leopold Fejer and Frederick Riesz, who were both pioneer researchers in the theory of Fourier series. Lillafüred, about a hundred miles from Budapest, was a resort with luxurious castle-like big hotels. Fejer and Riesz were in the habit of summering there. Fejer had been Johnny's teacher. Riesz was one of the most elegant mathematical writers in the world, known for his precise, concise, and clear expositions. He was one of the originators of the theory of function spaces—an analysis which is geometrical in nature. His book, *Functions of Real Variables,* is a classic. We were to walk in the forest, but in the morning before the walk Johnny said we had to wait until the master had his inspiration. By this he meant the daily physiological necessities—not spiritual ones, which had to be met with a pre-breakfast sip of brandy before the day could begin! We had a nice discussion. Of course the talk also concerned the world situation and the likelihood of war.

I returned to Poland by train from Lillafüred, traveling through the Carpathian foothills. I had to change trains several times and remember sitting for a while on an open flat car with my legs dangling over the side as we went through small villages named Satorolia-Ujhely and Munkaczewo. This whole region on both sides of the Carpathian Mountains, which was part of Hungary, Czechoslovakia, and Poland, was the home of many Jews. Johnny used to say that all the famous Jewish scientists, artists, and writers who emigrated from Hungary around the time of the first World War came, either directly or indirectly, from these little Carpathian communities, moving up to Budapest as their material conditions improved. The physicist I. I. Rabi was born in that region and brought to America as an infant. It will be left to historians of science to discover and explain the conditions which catalyzed the emergence of so many brilliant individuals from that area. Their names abound in the annals of mathematics and physics of today. Johnny used to say that it was a coincidence of some cultural factors which he could not make precise: an external pressure on the whole society of this part of Central Europe, a feeling of extreme insecurity in the individuals, and the necessity to produce the unusual or else face extinction. To me the picture was that of the Roman poet Virgil, describing the flood: "In the big whirlpool there appear only a few remaining swimmers," surviving through intellectual frenzy and strenuous and vigorous work. A jocular version of survival is a story I told to Johnny who manufactured many variations. A little Jewish farm boy named Moyshe Wasserpiss emigrated to Vienna and became a successful businessman. He changed his name to Herr Wasserman. Going on to Berlin and to even greater success and fortune, he became Herr Wasserstrahl, then von Wasserstrahl. Now in Paris and still more prosperous, he is Baron Maurice de la Fontaine.

Eugene Wigner is one of the famous scientists from Budapest. He and Johnny were school friends and studied together for a time in Zürich. Johnny told me a nice story from those days: Eugene and Johnny wanted to learn to play billiards. They went to a café where billiards were played and asked an expert waiter there if he would give them lessons. The waiter said, "Are you interested in your studies? Are you interested in girls? If you really want to learn billiards, you will have to give up both." Johnny and Wigner held a short consultation and decided that they could give up one or the other but not both. They did not learn to play billiards.

Von Neumann was primarily a mathematician. Wigner is primarily a physicist, but also half a mathematician and brilliant user of mathematics, a virtuoso of mathematical techniques in physics. I would add here that he recently published an interesting article on the a priori unexpected effectiveness of mathematics in physics. Von Neumann's book on the foundations of quantum theory had more philosophical and psychological meaning rather than direct applications in theoretical physics. Wigner made many concrete contributions to physics, perhaps nothing quite as overwhelming as Einstein's ideas of relativity, but many important specific technical achievements and also something rather general—namely, the very fundamental role of group theoretical principles in the physics of quantum theory and the physics of elementary particles.

When von Neumann died, Wigner wrote a beautiful obituary article in which he described the deep despair that came upon Johnny when he knew that he was dying, for it was impossible for him to imagine that he would stop thinking. For Wigner, von Neumann and thinking were synonymous.

After this visit to Budapest it was time to prepare my return to Harvard. I had to go to the American consulate in Warsaw each summer I was in Poland to apply for a new visitor's visa in order to return to the United States. Finally, the consul said to me, "Instead of coming here every summer for a new visa, why don't you get an immigration visa?" It was lucky that I did, for just a few months later these became almost impossible to obtain.

Twice I made plans to travel across the Atlantic with Johnny. The year his marriage with Marietta broke up, we went to Europe together. I joined him on the *Georgic*, a small Cunard boat which took a week to cross. Johnny always traveled first class, so I went first class also, although usually I traveled tourist. As always, we discussed a lot of mathematics. We flirted with a young woman named Flatau, whom we found rather attractive. The day after we met her, I asked him: "Have you solved the problem of Flatau?" He liked the play on words. In mathematics there is the famous problem of Plateau: given a curve or a wire in space, the problem is to find a surface such that the wire is its boundary and of minimal surface area. It can be demonstrated with soap bubbles. If you immerse a closed curved wire in a soap solution, you get some nice surfaces spanning them. The man who first formulated and studied this mathematically was Plateau.

In 1939, my three-year appointment at the Society of Fellows was expiring. Unfortunately, it was not renewable because I had passed the upper age limit. Thanks to G. D. Birkhoff, I received an extension of the stay at Harvard in the form of a lectureship in the mathematics department. More permanent prospects were not promising; it seemed there were no vacant assistant professorships. Because of the influx of large numbers of German and Central European scientists, and in spite of Johnny's efforts on my behalf, things did not look much better in Princeton. Thus, I returned to America with an assured position for only one more year, accompanied this time by my younger brother Adam, who was not quite seventeen. With Adam in my charge, a sense of responsibility came over me. Our mother had died the year before, and the feeling of impending crisis had convinced my father that Adam would be safer in

America with me. When he tried to apply for a visa, the American consul in Warsaw seemed to have reservations. It is only when I proved that I was living and teaching in the United States that he agreed to issue him a student visa.

Our father and uncle Szymon accompanied us to Gdynia, a Polish port on the Baltic Sea, to see us off on the Polish liner *Batory*. This was the last time we were to see either of them.

We were at sea when the announcement of the pact between Russia and Germany came over the ship's radio. In a state of strange agitation, I told Adam upon hearing the news, "This is the end of Poland." On a map in the ship's salon, I drew a line through the middle of Poland, saying, Cassandra-like, "It will be divided like that." We were, to say the least, shaken and worried.

At dinner on the first night, I suddenly noticed Alfred Tarski in the dining room. I had no idea he was on the boat. Tarski, famous logician and lecturer in Warsaw, told us he was on his way to a Congress on the Unity of Philosophy and Science to be held in Cambridge. It was his first trip to America. We ate at the same table and spent a good deal of time together. I still have an old shipboard photograph, which shows Adam, Tarski, and me dressed in dinner jackets ready for the gay American social life. He intended to stay only a couple of weeks and was traveling with a small suitcase of summer clothes. Because the war broke out shortly after we landed, he found himself stranded in the United States without money, without a job, and with his family—a wife and two small children—in Warsaw. For some time he was in the most precarious and terrible situation.

Adam was frightened and nervous when we landed, a young boy abroad for the first time and away from familiar surroundings. Johnny had come to meet us at the boat; on seeing Adam, he asked, out of his hearing, "Who is this fellow?" He had not heard my introduction and was surprised. There is a difference of thirteen years between my brother and me, and we do not look at all alike. Adam is taller than I, straight, blond, and with a pink complexion. I am somewhat dark and stockier. In appearance he resembles some of our uncles, while I look more like our mother. At the pier, Johnny appeared very agitated. People in the United States had a much clearer and more realistic view of events than we had had in Poland. For example, when it was time to obtain an exit visa from Poland, because I was in the Polish army reserve I had to first secure the permission of the army to leave the country. The officer in charge asked casually why I wanted to go abroad and raised no further question when I told him of my lecturing engagement in America. As a rule people in Poland had not felt the imminence of war, but rather a continuation of the state of crisis, similar to the one in Munich the year before.

We stayed in New York a few days, visiting cousins—the painter Zygmund Menkès and his wife—and a friend of the family who had given additional financial guarantees for Adam, who had a student visa. Actually, the plan was that my brother would receive monthly checks from home via our uncle's bank in England. We also saw young Mr. Loeb, an acquaintance of my cousin Andrzej, and when I talked to him on the telephone, he asked: "Will Poland give in?" I replied that I was pretty sure it would never surrender, and there was going to be a war.

I left Adam in New York to go with Johnny to Veblen's summer place in Maine. Though we were gone only two or three days, Adam was very unhappy with me for having left him. On the way to Veblen's, we discussed some mathematics as usual but mostly talked about what was going to happen in Europe. We were both nervous and worried; we examined all possible courses which a war could take, how it could start, when. And we drove back to New York. These were the last days of August.

Adam and I were staying in a hotel on Columbus Circle. It was a very hot, humid, New York night. I could not sleep very well. It must have been around one or two in the morning when the telephone rang. Dazed and perspiring, very uncomfortable, I picked up the receiver and the somber, throaty voice of my friend the topologist Witold Hurewicz began to recite the horrible tale of the start of war: "Warsaw has been bombed, the war has begun," he said. This is how I learned about the beginning of World War II. He kept describing what he had heard on the radio. I turned on my own. Adam was asleep; I did not wake him. There would be time to tell him the news in the morning. Our father and sister were in Poland, so were many other relatives. At that moment, I suddenly felt as if a curtain had fallen on my past life, cutting it off from my future. There has been a different color and meaning to everything ever since.

On the way to Cambridge, I accompanied Adam to Brown University in Providence, registered him as a freshman, and introduced him to a few of my friends, including Tamarkin and his son. His English was quite good, and he did not seem to mind being left alone in college.

I became a compulsive buyer of newspapers, all the extra editions every hour, and easily went through eight or ten papers a day looking for news of Lwów, of the military situation, and of the progress of battles. Early in September I saw in the Boston *Globe* a large photograph of Adam surrounded by other young freshmen at Brown. It was captioned, "Wonders whether his home was bombed."

From the start, Adam did very well in school; a few months later, he was able to obtain a tuition waiver. Nevertheless we were finding ourselves in severe financial straits. The proposed income from Britain had been frozen—the English government stopped all outgoing money, and my salary as a lecturer at Harvard was hardly sufficient to put a younger brother (who was not allowed to work because he was on a student visa) through college. On previous trips, I had never thought of transferring funds or property from Poland. Now it could no longer be done. I went to see a dean of the College and explained my situation. His name was Ferguson, and I feared that with such a Scottish name he would be rather parsimonious. Fortunately he was not. I told him that if I could not get a little more assistance from the University, I would be forced to leave my academic career and look for some other means of support. He was sympathetic and was able to find two or three hundred dollars more for the year, which was a sizable help in those days.

From indications at department meetings, I gathered that my chances of staying on at Harvard on a permanent basis were poor, so I began to make inquiries about another position for 1940. An assistant professorship became vacant at Lehigh University in Bethlehem, Pennsylvania, and I received a letter inviting me for an interview. I was not much interested in going to Bethlehem, but G. D. Birkhoff told me, "Stan, you must know that it is impossible in this country to get any advancement or raises in salary without offers from elsewhere. Yes, go have the interview at Lehigh." I replied, "Who will teach my class that day?" "I will," he said. I felt both embarrassed and honored that the great Professor Birkhoff would condescend to teach my class in undergraduate mechanics. Indeed, charmingly childish as he often was, and wanting to show the young students who he was, he gave a complicated and advanced lecture which, I later learned, they did not understand very well.

Bethlehem was covered with a yellow pall of acrid smog when I arrived for the interview—an inauspicious beginning. The chairman took me around the department and introduced me to a young professor there. It happened to be the number theorist D. H. Lehmer. As we entered his office, he was correcting a large pile of blue books and said to me, right in front of the chairman, "See what we have to do here!" This contributed to a negative impression and brought instantly to my mind a similar situation in Poland. When a chambermaid or some other servant girl was leaving a place of employment, she would take the new prospective maid aside and show her the less favorable aspects of the job.

This was the period of my life when I was perhaps in the

worst state, mentally, nervously, and materially. My world had collapsed. Prospects for the reconstitution of Poland in any recognizable form were dim indeed. There was a terrible anxiety about the fate of all those whom we had left behind—family and friends. Adam was also in a very depressed state, and this contributed to my worries. L. G. Henderson, who was always friendly and helpful, gave me all the moral support he could. When France collapsed in the spring of 1940, the situation became so dark and seemingly hopeless that despair gripped all the European émigrés on this side of the ocean. There was the added worry that should German ideas prevail along with their military successes, life in America would become quite different, and xenophobia and anti-Semitism might grow here too.

During that period I lived in a little room on the fourth floor of the Ambassador Hotel. On the fifth, the Alfred North Whiteheads lived in a large apartment whose walls were curiously painted in black. I knew Whitehead from the Society of Fellows dinners. He and Mrs. Whitehead held a weekly "at home" evening to which they invited me. He was already quite old, but his mind was crystal clear, sharp, incisive, with a better memory than many of his juniors. I remember their fortitude and courage at the time of the bombing of London. Whitehead never seemed to give up hope that the war would be won in the end, and he lived to see the defeat of Germany.

Conversation at the Whiteheads' was extremely varied. Besides the war, one discussed philosophy, science, literature, and people. The subject of Bertrand Russell came up once. He was having great troubles in this country. He had quarreled with Barnes, the Philadelphia millionaire who employed him at the time, and he was in difficulty at City College because of his views on sex and lectures on free love. Harvard tried to get him, but the invitation did not go through because of a wave of protests from proper Bostonians. I remember Whitehead talking about all this and Mrs. Whitehead exclaiming, "Oh, poor Bertie!" There was also talk of mathematics. Once someone asked, "Professor

Whitehead, which is more important: ideas or things?" "Why, I would say ideas about things," was his instant reply.

I spent much of my time with the other Poles who had found their way to Cambridge—Tarski, Stefan Bergman and Alexander Wundheiler. They were all terribly unhappy, Wundheiler most of all. He always had some kind of "Weltschmerz." We would sit in front of my little radio which I left on all day long, and listen to the war news. He would stay for hours in my room, and we drank brandy from toothbrush glasses. He was a talented mathematician, an extremely nice, pleasant, and intelligent person, with a mind rather hard to describe—that of an intelligent critic, but somewhat lacking the "it" of mathematical invention. Genius is not the word I mean. It is hard to describe the talent for innovation, even on a modest scale; besides, it exists in a continuous spectrum and is largely influenced by "luck." There may be such a thing as habitual luck. People who are said to be lucky at cards probably have certain hidden talents for those games in which skill plays a role. It is like hidden parameters in physics, this ability that does not surface and that I like to call "habitual luck." It is often remarked that in science there are people who have so much luck that one begins to feel it is something else. Wundheiler lacked this special spark.

I don't remember when and how he first appeared in the States. He had a temporary job at Tufts College in Boston. He had the usual impressions, complaints, appreciation and admiration for the United States; we talked about that at great length. Attitudes of students amused and shocked him. Being used to the more formal Polish manners, he was quite put out when in class one day a student called out to him, "Psst! The window is open. Can you close it?" One did not talk to professors that way in Poland.

He was interested in the geometry of the Dutch mathematician Schouten, which to me was too formal and symbolic. The notations were so complicated, I made fun of the formulae saying that they considered a geometric object a mere symbol, a letter around which indices were hung right, left, up, and down—like decorations on a Christmas tree.

We gradually lost contact after I left Cambridge. I learned later that he had committed suicide. I had a premonition of this because of a poem he would recite about a man who hanged himself with his tie. He was lonely, and many times he had told me of his unhappiness because of his looks. He was a very short man with an intelligent face, but not one that was considered appealing to women. He thought of himself as ugly, which bothered him.

In many cases, mathematics is an escape from reality. The mathematician finds his own monastic niche and happiness in pursuits that are disconnected from external affairs. Some practice it as if using a drug. Chess sometimes plays a similar role. In their unhappiness over the events of this world, some immerse themselves in a kind of self-sufficiency in mathematics. (Some have engaged in it for this reason alone.) Yet one cannot be sure that this is the sole reason; for others, mathematics is what they can do better than anything else.

Toward the end of the 1940 academic year Birkhoff intimated to me that there might be a vacancy at the University of Wisconsin. He added, "You should not be like the other European refugees who try at all costs to stay on the East Coast. Do as I have done, try to get a job in Madison. It is a good university; I was there as a young man." Taking his advice, I went to a meeting of the American Mathematical Society at Dartmouth to meet Professor Mark Ingraham, who was chairman of the mathematics department in Madison. In those days, mathematics meetings were job fairs at which one could see the department chairmen—almost like hereditary caciques, each surrounded by supplicants, groups of younger men looking for jobs. The situation was completely reversed in the late 1950s and early 1960s, when a lone young man fresh from school, with a brand-new Ph.D., would be surrounded by chairmen looking for young professors.

At the Dartmouth meeting I had a comical adventure. Late in the evening I walked into my dormitory room. It was dark, and I tried to get into bed without putting the light on. As I sat down, I heard a squeak and a groan. My bed had another occupant. I groped for the other bed. The occupant of my bed said, "Dr. Ulam?" I answered, "Yes." Immediately he said, "Given a group which is such and such, does it have this and that property?" I thought a moment and answered, "Yes," proceeding to outline a reason. "If it is compact, then it is true." "But if it isn't compact," he tried to continue. It was late, I was tired, and felt like saying, "If it isn't compact, to hell with it." I let the conversation drop and fell asleep.

G. D. Birkhoff seemed to like and appreciate my work. I think I know a possible reason. He liked my self-confidence and my near impudence in defending the point of view of modern mathematics based on set theory against his more classical approach. He admired the so-to-speak hormonal, emotional side of mathematical creativity. I probably reminded him of the way he felt when he was young. He liked the way I got almost furious when—in order to draw me out—he attacked his son Garrett's research on generalized algebras and more formal abstract studies of structures. I defended it violently. His smile told me that he was pleased that the worth and originality of his son's work was appreciated.

In discussing the general job situation, he would often make skeptical remarks about foreigners. I think he was afraid that his position as the unquestioned leader of American mathematics would be weakened by the presence of such luminaries as Hermann Weyl, Jacques Hadamard, and others. He was also afraid that the explosion of refugees from Europe would fill the important academic positions, at least on the Eastern seaboard. He was quoted as having said, "If American mathematicians don't watch out, they may become hewers of wood and carriers of water." He never said that to me, but he did often make slighting remarks about the originality of some foreigners. He also maintained that they ought to be content with more modest positions; objectively speaking, this was understandable and even fair. Just the same, I would sometimes get angry. Perhaps because my family had been so well off and that until 1940 I never had to give much thought to my financial situation, I could be independent and said what I thought openly. Once I countered one of his attacks on foreigners with, "What pleasure do you find in playing a game where the outcome does not depend on the skill of the opponent but on some external circumstance? What pleasure is there in winning a game of chess from a player who is forced to make poor moves because he needs help from his opponent?" He was quite taken back by this remark.

But Birkhoff helped me to secure the job in Madison. He spoke to Ingraham on my behalf; after the Dartmouth meeting, I received an offer of an instructorship in Madison. At thirty, and with a certain reputation among mathematicians in Poland and America, I felt I could have been offered at least an assistant professorship. But the circumstances were so far from normal, and with someone like Jacques Hadamard, the most celebrated French mathematician, having been offered a lectureship in New York, or Tarski having been taken on as an instructor in Berkeley, I swallowed my pride and accepted the offer. Financially, it was not bad—something like twenty-three hundred dollars a year. Nevertheless, I felt sad at leaving Harvard and the "cultural East" for what I believed to be a more primitive and intellectually barren Middle West. On the East Coast it was implied that Harvard and perhaps Yale and Princeton were the only places with "culture." I was sure that Madison, about which I knew nothing, would be like Siberia and that I was being exiled. But since there were no alternatives, I prepared to leave Cambridge at the end of the summer, grimly determined to live through the years of exile and to await the outcome of the war.

CHAPTER 7

The University of Wisconsin

1941–1943

I TRAVELED to Madison by way of Chicago, where I changed to a smaller train that made several whistle stops, one at a town named Harvard. The irony did not escape me, and I felt fate was playing a cruel joke on me. It did not take long for me to change my outlook completely. I promptly found that the state of Wisconsin had important liberal political traditions, that the famous La Follettes had left their imprint not only on the state capital but on the University as well. The entire physical impression, the landscape, the lakes, the woods, the houses, and the size of the city were most agreeable. Living conditions were a pleasant surprise. I was given a room at the University Club and almost at once met there congenial, intelligent people not only in mathematics and science, but also in the humanities and arts. The rooms were small, with bath, bed, desk, and chairs. (I remembered how in one of Anatole France's novels, the hero, Father Coignard, says, "All one needs is a table and a bed. A table on which to have in turn learned books and delicious meals, a bed for sweet repose and ferocious love!") Downstairs there were comfortable common rooms, a library, dining room, even a game room with billiard tables.

The university was adequately supported by state funds and had extra income from a former professor's discovery of a special treatment of milk, the patent rights for which belonged to the University.

Johnny's friend Eugene Wigner was a physics professor there. I had a letter of introduction to another eminent physicist, Gregory Breit, from Harlow Shapley, the Harvard astronomer with whom I had had pleasant scientific and social contacts during my stay in Cambridge. It was Shapley who discovered the "scale of the universe" by using the luminosity period of the cepheid stars as markers. I quickly made friends with most of the mathematicians—many of them my contemporaries, Steve Kleene the logician, C. J. Everett, Donald Hyers, and others. Being by nature gregarious, I liked living at the University Club, meeting and taking meals with interesting colleagues.

One of them was Vassilief, a Russian émigré, a great expert on Byzantine history, and almost a character out of Nabokov's book *Pnin*. At dinner, he always ordered a second bowl of soup and would say to me, "Americans are funny; even when the soup is excellent, they never think of ordering a second bowl." Like many Russians, he liked to drink and carried a small flask of vodka in his coat pocket. He must have been in his sixties at the time. Some two years later, when the U.S. Army took over the Faculty Club for its quarters, Vassilief and the other occupants had to find other housing. Vassilief was given a two-room suite in a private house. He was enchanted with this new spaciousness. "It's wonderful," he explained. "You can sleep in one room and work in the other." And just like Pnin he threw what he called "a house-heating party to celebrate."

Another interesting person was a professor of English literature, a bachelor, Professor Henley. Thanks to my memory, which enabled me to quote Latin and to discuss Greek and Roman civilization, it became obvious to some of my colleagues in other fields that I was interested in things outside mathematics. This led quickly to very pleasant relationships. Henley was a good billiard player. He insisted on teaching me how to play, though he was rather appalled by my ineptitude. This, I found, is a very nice and very American trait—the desire to coach and instruct.

So I found Madison not at all the intellectual desert I had feared it would be. The university had a tradition of excellence in several fields of the natural sciences. It had great expertise in limnology. Limnology, the science of lakes, was developed by an old professor whose name I cannot recall, but who used to say, I was told, that every time he remembered the name of a student he forgot the name of a fish. Biology, too, was strong at the University of Wisconsin, as well as economics and political science. The economist Selig Perlman, and Nathan Feinsinger, who later became nationally known as a labor relations expert, were there at the time among many other eminent professors.

It also seemed that foreigners like myself who were, so to speak, not unpresentable, were well received into the academic community's social life and quickly established good relations with many professors in various fields. The intellectual atmosphere was lively. On the whole the professors did not put on airs as a few have at Harvard. On the contrary, perhaps in order to bear comparison with the famous older universities, they worked more energetically, but the "Oh, excuse me I've got to get to work" syndrome was not as evident as at Harvard.

Something else happened to make Madison most important to me. It was there that I married a French girl, who was an exchange student at Mount Holyoke College and whom I had met in Cambridge, Françoise Aron. Marriage, of course, changed my way of life, greatly influencing my daily mode of work, my outlook on the world, and my plans for the future.

The poet William Ellery Leonard, a tall man with a very large head and a mane of white hair, was one of the interesting and colorful members of the faculty, a great eccentric. The author of the book *The Locomotive God*, he was reputed to have an intense, neurotic fear of trains. This prevented him from ever leaving the University in Madison; the story was that his salary (which was very low for a full professor) never was increased, because he would never leave anyway. I found this reason rather humorous.

At that time, at many universities deans and chairmen often ran their departments not so much for excellence in scholarly or educational pursuits, but for good economy and efficiency—rather like a business. Indeed, shortly after I arrived, someone pointed out that the marvelous physical location of the campus on the shores of Lake Mendota constituted a part of our salaries, making them a little lower than at other comparable state universities. This made me joke with some of my younger colleagues: every time we looked at the beautiful lake it was costing us about two dollars. At one of the first faculty meetings I attended, Clarence A. Dykstra, the President, very imposing in appearance (and actually a very good man), started his speech with, "All of us face a challenge this year." At this I nudged my neighbor and whispered, "Watch out! This means no raises for the faculty." Sure enough, ten minutes later Dykstra said something to this effect, and my neighbor laughed out loud.

The astronomer Joel Stebbins was a professor in Madison. I enjoyed meeting him and talking to him in the observatory. He had an excellent sense of humor and was a great practical joker. Once on a clear, cold sunny winter Sunday, he drove to our apartment and honked the car horn. I looked out and there he was, saying, "Would you like to come with me to the Yerkes Observatory? There is a meeting of the Astronomical Society there." Yerkes was not far from Madison—about a two-hour drive. I dressed warmly, hurried down and on the way we discussed all kinds of problems. Then teasingly he said: "Would you like to give a talk?" To answer a joke with a joke, I responded, "Yes, for five or ten minutes." Quickly I started to think about what I could say to astronomers in a few minutes. I remembered that once I had been thinking about the mathematics of the way trajectories of celestial bodies might look from a moving system of coordinates and how, by a suitable motion of the observer, one could make complicated-looking orbits appear simpler if one assumed that the observer was himself moving. I called this general question "The Copernican Problem" and spoke for a few minutes on that. Indeed, it is a worthwhile subject to consider, and it does give rise to some bona fide topological and metric questions in which I had obtained a few simple results.

From the first year of my employment I was given a very light teaching load—namely only eleven hours of elementary courses (recognizing the fact that I was doing research and writing a number of papers), while some of the other instructors taught thirteen or sixteen. Later, this was reduced to nine hours per week. These elementary courses required essentially no preparation on my part, aside from an occasional look at the sequence of topics in the book so that I would cover the prescribed material and not go too quickly or too slowly. But the very expression "teaching load" as used by almost everyone from famous scholars to administrators was not only repugnant to me but ridiculous. It implied physical effort and fatigue—two things I have always been afraid of, lest they interfere with my own thinking and research. I was grateful to Ingraham, the head of the department, for understanding this. He was a jovial and pleasant man, who used to come to the faculty club on weekends to watch movies of football games. Known for his fondness for apple pie with cheese, he introduced me to the cheeses of Wisconsin, one of the state's dairy specialties, before I made a later acquaintance with the infinite variety of those of France.

In general, teaching mathematics is different from teaching other subjects. I feel, as do most mathematicians, that one can teach mathematics without much preparation, since it is a subject in which one thing leads to another almost inevitably. In my own lectures to more advanced audiences, seminars, and societies, I discuss topics that currently preoccupy me; this is more of a stream of consciousness approach.

I am told that I teach rather well. This is possibly because I believe one should concentrate on the essence of the subject and also not teach all items at a uniform level. I like to stress the important and, for contrast, a few unimportant details. One remembers a proof by recalling, as it were, a sequence of pleasant and unpleasant points—that is, difficult and easy ones. First comes a difficulty on which one makes an effort, then things go automatically for a while, suddenly again there is a new special trick that one has to remember. It is somewhat like going through a labyrinth and trying to remember the turns.

When I was teaching calculus in Madison (and it is a marvelous thing to teach) and had worked out a problem on the board, I was amused when some student would say, "Do another one like it!" They didn't even have a name for "it." Needless to say, these students did not become professional mathematicians.

One may wonder whether teaching mathematics really makes much sense. If one has to explain things repeatedly to somebody and assist him constantly, chances are he is not cut out to do much in mathematics. On the other hand, if a student is good, he does not really need a teacher except as a model and perhaps to influence his tastes. A priori, I tend to be pessimistic about students, even the bright ones (though I remember some good students at Harvard with whom I could talk and feel that teaching was not merely an empty exercise).

Generally speaking, I don't mind teaching as such, although I do not like to spend too much time at it. What I dislike is the obligation to be at a given place at a given time—not being able to feel completely free. This is because one of my characteristics is a special kind of impatience. When I have a fixed date, even a pleasant dinner or party, I fret. And yet when I am completely free, I may become restless, not knowing what to do.

With my friend Gian-Carlo Rota I once computed that including seminars and talks to advanced audiences, we must have lectured for several thousand hours in our lives. If one recognizes that the average working year in this country amounts to around two thousand hours, this is a large proportion of one's waking time, even if it is not completely "waking" since teaching is sometimes done in a partial trancelike state.

It was in Madison that I met C. J. Everett, who was to become my close collaborator and good friend. Everett and I hit it off immediately. As a young man he was already eccentric, original, with an exquisite sense of humor, wry, concise, and caustic in his observations. He was totally devoted to mathematics—they were his only interest. I found in him much that resembled my friend Mazur in Poland, the same kind of epigrammatic comments and jokes. Physically, they had a certain similarity, both being thin, bony, and less than medium height. Even their handwriting was similar; they both wrote in very neat, almost microscopic little symbols. Everett is several years younger than I. We collaborated on difficult problems of "order"—the idea of order for elements in a group. In our mathematical conversations, as always, I was the optimist, and had some general, sometimes only vague ideas. He supplied the rigor, the ingenuities in the details of the proof, and the final constructions.

One paper we wrote on ordered groups caught the fancy of the woman who was head of one of the women's military organizations during the war. At a meeting we heard her describe the activities of the corps by calling the organization "ordered groups."

Later we wrote another joint paper on projective algebras. I think this was the first attempt to algebraize mathematical logic beyond the so-called Boolean or Aristotelian elementary operations to include the operations "there exists" and "for all," which are both vital and comprehensive for advanced mathematics.

We both taught courses for naval recruits in 1942 and 1943. Also, in order to earn some extra money, we corrected papers for the Army Correspondence School. Françoise helped in that, too—she could do it very well for elementary arithmetic and algebra exercises. The Correspondence School paid thirty-five cents per paper; this amounted to quite a bit of money and began to reach sums comparable to the university salary. At this point, the administration decided to step in and impose restrictions on the number of papers one person could correct. The Army correspondence work was administered by an older woman who was a member of the mathematics department; it was supervised by a professor, Herbert Evans, a very jovial and pleasant person with whom I became friendly. He was one of the most good-humored persons I have ever known anywhere.

Everett and I shared an office in North Hall, an old structure halfway up the hill which housed the mathematics department. Leon Cohen, a visiting professor from the University of Kentucky with whom we had published some joint work, was there with us. We spent hours in that office, the three of us; the entire building would resound with our frequent laughter. Before and after classes, we corrected student papers—an occupation I hated and always tried to put off. As a result my desk was piled high with stacks of uncorrected workbooks and as I deposited new ones at one end, the older ones at the other mercifully dropped into the waste basket. Sometimes the poor students wondered why I was not returning their work.

After lunch, we played billiards—or tried to. Henley's lessons at the Faculty Club had very little effect on my game. Fun in this North Hall office and our frequent sessions in the Student Union, a luxurious building on the shore of the lake, were among the charms of life in Madison. This combination of leisure plus informal stimulation plays an important role in one's mental activity. Beyond the merely agreeable physical setting, it is often more valuable than the more formal gatherings at seminars and meetings which lead to discussions of a drier type. In its way, this somewhat replaced for me the old sessions in the coffee houses of Lwów for which I have had a longing ever since leaving Poland.

Everett stayed in Madison throughout the war. Later he joined me at Los Alamos, where we did much more work together including our now well-known collaboration and work on the H-bomb.

Everett exhibited a trait of mind whose effects are, so to speak, non-additive: persistence in thinking. Thinking continuously or almost continuously for an hour, is at least for me—and I think for many mathematicians—more effective than doing it in two half-hour periods. It is like climbing a slippery slope. If one stops, one tends to slide back. Both Everett and Erdös have this characteristic of long-distance stamina.

There were also Donald Hyers and Dorothy Bernstein. Hyers also had persistence in thinking about problems and an ability to continue to push the train of thought on a specific problem; we wrote several joint papers together. Dorothy Bernstein was a graduate student in my class. She was an interested, enthusiastic, and faithful taker of notes and organizer of material from a course I taught on measure theory. She collected much material, and we intended to write a book together, but our work was interrupted by my departure from Madison in 1944 and our plans were abandoned.

One day in my office, a young and brilliant graduate student named Richard Bellman appeared and expressed a desire to work with me. We discussed not only mathematics but the methodology of science. When the United States entered the war, he wanted to go back East—I believe to New York where he came from—and asked me to help him obtain a fellowship or a stipend so he could continue working after he left Madison. I remembered that in Princeton Lefschetz had some new scientifico-technological enterprise connected with the war effort. I wrote to him about Bellman in a sort of Machiavellian way, saying that I had a very able student who was so good that he deserved considerable financial support. I added that I doubted that Princeton could afford it. This, of course, immediately challenged Lefschetz and he offered Bellman a position. Two years later, Dick Bellman appeared suddenly in Los Alamos in uniform as a member of the SEDs, a special engineering detachment of bright and scientifically talented draftees who had been assigned to help with the technical work.

Through my connections with physicists and a seminar I taught in physics in the absence of Gregory Breit, I heard of the recent arrival in Madison of a very famous French physicist, Léon Brillouin. I called on him and found that his wife, Stéfa, was Polish, born in the city of Lódz, a large textile-manufacturing town. Stéfa and Léon had met when she was a young student in Paris, and they were married sometime before World War I. When World War II broke out he was, I think, a director of the French broadcasting service, with all the military responsibilities that entailed. After the collapse of France and the installation of the Vichy régime, he managed to escape from France at the first opportunity. He was internationally known for his work in quantum theory, statistical mechanics, and also in solid state physics. In fact, he was one of the pioneers in the theory of solid state. (The idea of "Brillouin" zones and other important notions are due to him.) He was also a very prolific and successful writer of physics textbooks and monographs.

Mrs. Brillouin had a great artistic flair. In the early nineteen-twenties she acquired, for modest sums, many works by Modigliani, Utrillo, Vlaminck, and other painters. In Madison she herself started to paint flower compositions in oil made up of thick layers—a style all her own. The Brillouins invited us to stop at their apartment the day

Françoise and I were married. They held a small surprise reception for us, where we drank French champagne and partook of a memorable cake by Stéfa. Stéfa Brillouin spoke hardly any English, but a few weeks after her arrival, in shopping for various objects she discovered that "le centimètre d'ici" as she called our inch, was about two and a half times the "centimètre de France." This almost exact estimate, the inch being 2.54 centimeters, was obtained solely by looking at the sizes of materials, curtains, and rugs. In Madison a close association began, which continued long after the war and lasted until their deaths a few years ago.

Before my second year in Madison, I was promoted to an assistant professorship—a step which gave me hope and some confidence in the material aspects of the future. To start a home and at the same time help support my brother on my modest salary (of twenty-six hundred dollars a year) was difficult. Often to make ends meet I visited the Faculty Credit Union, where a sympathetic officer made me loans of up to one hundred dollars, which had to be repaid in a few months' time.

I was asked to run the mathematics colloquium, which took place every two weeks and involved both local and visiting mathematicians. I might add that the payments to speakers were ridiculously small; even for those times, they amounted to about twenty-five dollars—and this included traveling expenses.

The colloquium was run differently from what I had known in Poland, where speakers gave ten- or twenty-minute informal talks. At Madison they were one-hour lectures. There is quite a difference between short seminar talks like those at our math society in Lwów, and the type of lecture which necessitates talking about major efforts. The latter were better prepared, of course, but their greater formality removed some of the spontaneity and stimulation of the shorter exchanges. In this connection I met André Weil, the talented French mathematician, who had gone to South America at the beginning of the war. He disliked conditions there and came to the United States, where he had gotten a job at Lehigh. Weil was already well known internationally for his important work in algebraic geometry and in general algebra. His colloquium talk was on one of his most important results on the Rieman hypothesis for fields of finite characteristic. The Rieman hypothesis is a statement that is not easy to explain to laymen. It is important because of its numerous applications in number theory. It has challenged many of the greatest mathematicians for about a hundred years. It is still unproved although considerable progress has been made toward a possible solution.

Dean Montgomery, a friend I had met at Harvard, came at my invitation and gave a colloquium talk. There was a vacancy in the department, and I tried to interest him in coming to our university, where Ingraham and Langer, the two most senior professors, were both very much in favor of his appointment; instead he went to Yale. Later he told me stories about the atmosphere at Yale, which at that time in some circles was ultra-conservative. In his interview he was asked whether he was for or against Jews in the academic profession and also whether he was a liberal. Even though he answered both questions "wrongly" from the interlocutor's point of view, he nevertheless received an offer. He left Yale a few years later to join the Institute in Princeton.

Eilenberg and Erdös were also among the speakers invited to the colloquia. Erdös was one of the few mathematicians younger than I at that stage of my life. He had been a true child prodigy, publishing his first results at the age of eighteen in number theory and in combinatorial analysis.

Being Jewish he had to leave Hungary, and as it turned out, this saved his life. In 1941 he was twenty-seven years old, homesick, unhappy, and constantly worried about the fate of his mother who had remained in Hungary.

His visit to Madison became the beginning of our long, intense—albeit intermittent—friendship. Being hard-up financially—"poor," as he used to say—he tended to extend, his visits to the limits of welcome. By 1943, he had a fellow-ship at Purdue and was no longer entirely penniless—"even out of debts," as he called it. During this visit and a sub-sequent one, we did an enormous amount of work together—our mathematical discussions being interrupted only by reading newspapers and listening to radio accounts of the war or political analyses. Before going to Purdue, he was at the Institute in Princeton for more than a year with only a pittance for support that was to end later.

Erdös is somewhat below medium height, an extremely nervous and agitated person. At that time he was even more in perpetual motion than now—almost constantly jumping up and down or flapping his arms. His eyes indicated he was always thinking about mathematics, a process inter-rupted only by his rather pessimistic statements on world af-fairs, politics, or human affairs in general, which he viewed darkly. If some amusing thought occurred to him, he would jump up, flap his hands, and sit down again. In the intensity of his devotion to mathematics and constant thinking about problems, he was like some of my Polish friends—if pos-sible, even more so. His peculiarities are so numerous it is im-possible to describe them all. One of them was (and still is) his own very peculiar language. Such expressions as "epsi-lon" meaning a child, "slave" and "boss" for husband and wife, "capture" for marriage, "preaching" for lecturing, and a host of others are now well known throughout the mathe-matical world. Of all the results we obtained jointly, many have still not been published to this day.

Erdös has not changed much as the years have gone by. He is still completely absorbed by mathematics and mathe-maticians. Now over sixty, he has more than seven hundred papers to his credit. Among the many sayings about him one goes, "You are not a real mathematician if you don't know Paul Erdös." There is also the well-known Erdös number —the number of steps it takes any mathematician to connect with Erdös in a chain of collaborators. "Number two," for example, is to have a joint paper with someone who has written a paper with Erdös. Most mathematicians can usually find a link with him, if not in one then in two stages.

Erdös continues to write short handwritten letters beginning, "Suppose that x is thus and so . . ." or "Suppose I have a sequence of numbers . . ." Toward the end, he adds a few personal remarks, mainly about getting old (this started when he was thirty) or with hypochondriac or pessimistic observations about the fate of our aging friends. His letters are nevertheless charming and always contain new mathematical information. Our correspondence started before the Madison years, containing many discussions of the hardships of young mathematicians who could not find jobs or of how to deal with officials and administrators. He used the expression, "Oh, he is a big shot" about young American assistant professors, and when he called me one, I introduced a subclassification of "big shots, small shots, big fry and small fry"—four orders of status. In 1941, as an assistant professor I told him I was at best a "small shot." This amused him, and he would allot one of these four grades to our friends in conversations or correspondence.

Through thick and thin, Johnny von Neumann and I also continued our correspondence, which included a little mathematics even in those days, and a lot about the tragic happenings in the world. There was much isolationism in the United States, and the obvious and widespread disinclination to enter the war created in me a feeling of despair mixed with resentment. On the whole, Johnny was more optimistic and knew better than I the power of the United States and the long-range goals of U.S. policies. He was already an American citizen engaged (though I did not know it at the time) in the war effort the country was preparing.

The tone of our mathematical correspondence and of our conversations when we met at mathematics meetings changed from the abstract to more applied, physics-related topics. He was now writing about problems of turbulence in hydrodynamics, aerodynamics, shocks, and explosives.

Johnny held discussions with many scientists, among them Norbert Wiener. Although Norbert was a pacifist, he badly wanted to contribute something important to the American war effort. Wiener felt as Russell did that this was a "just war," a necessary war, and the only hope for mankind lay in U.S. intervention and victory. But Norbert was difficult in his dealings with the military, whereas Johnny always got along with them.

Wiener wrote in his autobiography that he had ideas similar to the ones I later proposed as the Monte Carlo method. He says vaguely that he found no response when he talked to someone and so dropped the matter, in the same way that he lost interest in the idea of geometry of vector spaces and function spaces à la Banach. In fact, in one of his books he called these vector spaces (which are associated with Banach's name alone) Banach-Wiener spaces. This nomenclature did not "take" at all.

In the first World War mathematicians had done much work in classical mechanics, calculations of trajectories, and external and internal ballistics. This work was resumed at the beginning of the second World War, although it soon turned out not to be the main thrust of the scientific applications. Hydrodynamic and aerodynamic questions became more detailed and urgent, particularly because they were directly connected with special war problems. Early in 1940 I took from the library a German textbook on ballistics and studied it, but noticed that there was not much in it of importance to the military technology of the forties. At the beginning of the war electronic computing machines did not exist. There were only the beginnings of the mechanical relay machines constructed at Harvard, at IBM, and one or two other places.

As soon as the prescribed time had elapsed I applied for and received American citizenship in Madison in 1941. I hoped that this would make it easier for me to enroll in the war effort. To pass the examination I studied the history of the United States, the essence of the Constitution, the names of Presidents, and the other topics one was likely to be asked about at the examination. I don't remember why, but instead of my having to go to Chicago, an examiner came to Madison to our apartment. After a few words I noticed that he must have been an immigrant himself or the son of an immigrant. His appearance was quite Jewish, and perhaps impudently I asked him about his own origins and background. He did not seem to mind and replied that his parents had come from the Ukraine. Soon I realized with embarrassment that it was I who was examining him.

Immediately after I received the citizenship papers, I tried to volunteer in the Air Force. At thirty I was considered too old to become a combat pilot, but with my mathematical background I thought I could receive training as a navigator, because the University had received a notice that the Air Force was looking for volunteers. I went to a recruiting center not far from Madison for a physical examination. It was given by West Coast Japanese medics who had been relocated in this Middle Western camp. Because the physical tests involved the taking of blood samples, I told myself jokingly that I was losing some blood to the Japanese in defense of my new country. I was disappointed when my application was rejected because of my peculiar eyesight.

Teaching army math courses did not seem sufficiently relevant; I wanted to do something more immediately useful, something that would contribute more directly. I thought of going to Canada to enlist there, and I remembered a conversation in Cambridge in 1940 with Whitehead, who had relatives who were officers in the Royal Canadian Air Force. So I wrote him asking whether he could help me to become involved in the war effort in Canada. He sent back a letter which I treasure for all the things it said. Even though he said that he had written to someone in Canada on my behalf, nothing came of this.

Then Los Alamos entered the picture.

S. M. Ulam, 1938. Pencil sketch by Zygmund Menkès

With fellow mathematician Stanislaw Mazur (*left*), Lwów, ca 1933

Stefan Banach, Poland, ca 1945

Polish mathematics students convention, Lwów, 1930
Designated by numbers: (1) Leon Chwistek, (2) Stefan Banach, (3) Stanislaw
Loria, (4) Kazimir Kuratowski, (5) Stefan Kaczmarz, (6) Juliusz Schauder,

(7) Marceli Stark, (8) Karol Borsuk, (9) Edward Marczewski,
(10) S. M. Ulam, (11) A. Zawadzki, (12) Edward Otto, (13) W. Zonn,
(14) M. Puchalik, (15) K. Szpunar

On the dock at Gdynia before Stan and his younger brother, Adam, embarked for America in 1939. The four Ulams are (*left to right*) Szymon (uncle), Adam, Jozef (father), and Stan

John von Neumann, Princeton, 1932

The Harvard Junior Fellows, Cambridge, 1938
Left to right, seated: George Homans, Jim Fisk, Paul Samuelson, John Snyder, James Miller, Ivan Getting, Willard Quine, Robert Woodward, George Hass
Standing, first row: James Baker, Kenneth Murdock, Paul Ward, George Haskins, L. J. Henderson, John Ferry, George Hanfmann, Charles Curtiss, Alfred North Whitehead, John Livingston Lowes, Talbot Waterman, Tom

Chambers, Samuel Eliot Morison, John Miller, Conrad Arensberg, David
Griggs, William Whyte
Back row: F. Edward Cranz, Reed Rollins, Harry Levin, Frederick
Watkins, John Oxtoby, E. Bright Wilson, Richard Howard, Albert Lord,
Garrett Birkhoff, Craig LaDrière, Stan Ulam, Orville Bailey
(Harvard University)

Ulam and C. J. Everett in front of North Hall at the University of Wisconsin, Madison, 1941

The entrance to wartime Los Alamos (Los Alamos Scientific Laboratory)

Enrico Fermi in the 1940s (Harold Agnew)

Nobel prizewinners Ernest O. Lawrence, Fermi, and I. I. Rabi, during a meeting at the Lodge, Los Alamos, in the late 1940s

Lunch at the Lodge; (*clockwise from right*) Richard Feynman, Carson Mark, Jack Clark, Fermi

Von Neumann, Feynman, and Ulam on the porch of Bandelier Lodge in Frijoles Canyon, New Mexico, during a picnic, ca 1949 (Nicholas Metropolis)

Tennis at Los Alamos, 1958; (*left to right*) James Tuck, Ulam, Conrad Longmire, Donald Dodder (Los Alamos Scientific Laboratory)

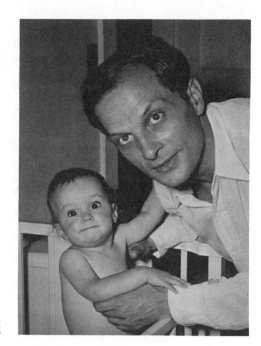

Claire Ulam, aged one,
with her father, 1945

Ulam standing on the piece of land in Santa Fe that he bought for $150 in
1947

The von Neumanns starting the descent into the Grand Canyon on an excursion in the late 1940s: Klari, with visor, is fourth from front; Johnny, bareheaded and in city suit, is last, on the only mule facing the wrong way

On the Plaza in Santa Fe, ca 1949, von Neumann and his thirteen-year-old daughter, Marina

Johnny, Claire, and Stan in the Ulam yard at Los Alamos, ca 1954

George Gamow's representation of the "super" directing committee; (*bottom; left to right*): Ulam, Edward Teller, Gamow; Joseph Stalin and J. Robert Oppenheimer observe proceedings from above (The Viking Press)

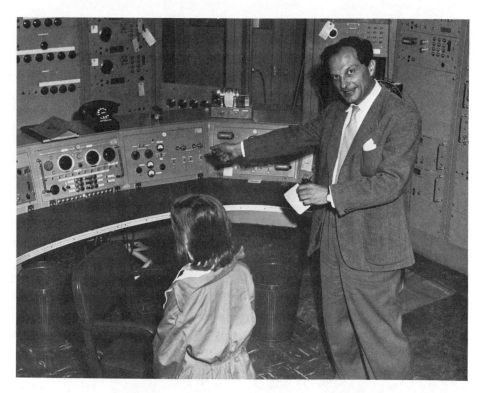

"Demonstrating" the MANIAC to Claire at a laboratory open house, 1955 (Los Alamos Scientific Laboratory)

Ulam foiled by the technological problem of fitting Theodore von Kármán with the neck cord of a microphone (Max Spring, Los Alamos Scientific Laboratory)

An informal moment during a presidential visit, Los Alamos, 1962; *(left to right)* Vice President Lyndon B. Johnson, President John F. Kennedy, Senator Clinton P. Anderson, mathematician Ulam, Congressman Joseph M. Montoya, science advisor Jerry Wiesner (Los Alamos Scientific Laboratory)

Ulam and *Life* Magazine photographer Alfred Eisenstaedt near the Los Alamos laboratory, 1962 (William H. Regan, Los Alamos Scientific Laboratory)

Stan and Françoise Ulam at home in Los Alamos, 1964 (Lloyd Shearer)

PART III

Life
among
the Physicists

CHAPTER 8

Los Alamos

1943–1945

DURING the late spring of 1943, I wrote to von Neumann about the possibility of war work. I knew he was involved, because his letters often came from Washington rather than from Princeton. I was not happy with teaching, although I did a lot of mathematics, wrote papers, organized colloquia, and taught war-related courses. Still it seemed a waste of my time; I felt I could do more for the war effort.

One day Johnny answered with an intimation that there was interesting work going on—he could not tell me where. From Princeton he said he was going west via Chicago, and suggested that I come to the Union Station there to talk to him since he had two hours between trains. That was in the early fall of 1943.

I went and, sure enough, Johnny appeared. What caught my attention were the two men escorting him, looking a bit like "gorillas." They were obviously guards, and that impressed me; he must be an important figure to rate this, I decided. One of the men went to do something about his railroad ticket, and we talked in the meantime.

Johnny said that there was some very exciting work going on in which I could possibly be of good use; he still could not tell me where it was taking place, but he traveled rather often from Princeton to that location.

I don't know why—by pure chance or one of these incredible coincidences or prophetic insights?—but I answered jokingly, "Well, as you know, Johnny, I don't know much about engineering or experimental physics, in fact I don't even know how the toilet flusher works, except that it is a sort of autocatalytic effect." At this I saw him wince and his expression become quizzical. Only later did I discover that indeed the word autocatalytic was used in connection with schemes for the construction of an atomic bomb.

Then another coincidence occurred. I said, "Recently I have been looking at some work on branching processes." There was a paper by a Swedish mathematician about processes in which particles multiply quite like bacteria, for example. It was prewar work, and an elegant theory of probabilistic processes. That, too, could have had something to do with the mathematics of neutron multiplication. And again he looked at me almost with suspicion or wonder and smiled wanly.

The Wisconsin astronomer Joel Stebbins, whom I saw occasionally, had told me about some work going on with uranium and about the release of energy from very heavy elements. I wondered if, subconsciously, this prompted my remarks.

During this meeting at the station, Johnny and I also discussed what seemed a general lack of imagination in the scientific community's planning of work useful for the war effort—especially in computations for hydrodynamics and aerodynamics. I pointed out my doubts about the age of some of the main participants (people over forty-five seemed to me at that time old). Johnny agreed there were obvious elements of senility. As usual, we tried to lighten our sadness with jocular comments, observing that someone should establish a "gerontological" society, whose members would be scientists interested in war work and afflicted with premature or "galloping" senility.

Since Johnny could not or would not tell me where he was going except that it was to the Southwest, I remembered an old Jewish story about two Jews on a train in Russia. One asks the other, "Where are you going?" and the second replies, "To Kiev." Whereupon the first says, "You liar, you tell me you are going to Kiev so I would think you are going to Odessa. But I know you are going to Kiev, so why do you lie?" And I told Johnny, "I know you can't tell me, but you say you are going Southwest in order that I should think that you are going Northeast. But I know you are going Southwest, so why do you lie?" He laughed. We talked a while longer about the war situation, politics, and the world; then his two companions reappeared and he left.

I saw him once more, I think, in Chicago, before I received an official invitation to join an unidentified project that was doing important work, the physics having something to do with the interior of stars. The letter inviting me was signed by the famous physicist Hans Bethe. It came together with a letter from the personnel department with details of the appointment, salary, clearance procedures, indications on how to get there, and so forth. I accepted immediately with excitement and eagerness.

The pay was slightly above my university salary, but on a twelve-month basis—around $5,000, if I remember correctly. The professional physicists like Bethe who were there already received little more than their university salaries. I learned later that a chemist from Harvard, George Kistiakowski, had the allegedly astronomical salary of $9,000 or $10,000.

I informed my university of this opportunity to join an obviously important war project and secured a leave of absence for the duration.

A student of mine, Joan Hinton, had left for an unknown destination a few weeks before. Joan was taking a course I gave in classical mechanics. One day she appeared in my office in North Hall to ask if I could give her an examination three or four weeks before the end of the term so that she could start some war work. She produced a letter from Professor Ingraham, the chairman, authorizing me to do that. She was a good student, a rather eccentric girl, blonde, sturdy, good-looking. Her uncle was G. I. Taylor, the English physicist. She was also a great-granddaughter of George Boole, the famous nineteenth-century logician. I wrote a number of questions on the back of an envelope; Joan took some sheets of paper, sat down on the floor with her notebook, wrote out her exam, passed, and disappeared from Madison.

Soon after, other people I knew well began to vanish one after the other, without saying where—cafeteria acquaintances, young physics professors and graduate students like David Frisch, and his wife Rose, who was a graduate student in my calculus class, Joseph McKibben, Dick Taschek, and others.

Finally I learned that we were going to New Mexico, to a place not far from Santa Fe. Never having heard about New Mexico, I went to the library and borrowed the Federal Writers' Project Guide to New Mexico. At the back of the book, on the slip of paper on which borrowers signed their names, I read the names of Joan Hinton, David Frisch, Joseph McKibben, and all the other people who had been mysteriously disappearing to hush-hush war jobs without saying where. I had uncovered their destination in a simple and unexpected fashion. It is next to impossible to maintain absolute secrecy and security in war time.

This reminds me of another story. Since I knew Stebbins well, about a month after arriving at Los Alamos, I wrote to him. I did not say where I was but mentioned that in January or February I had seen the star Canopus on the horizon. Later it occurred to me that as an astronomer he could easily have deduced my latitude since this star of the Southern skies is not visible above the 38th parallel.

I shall pass over our problems in getting train reservations. Even with the priorities I had, our departure was delayed by about a month. On the train I had to offer a gratuity to the conductor to obtain a berth for Françoise, who was two months pregnant at the time. This was the first—and I think last—time in my life that I "bribed" anyone.

We arrived at a remote, lonely, unimpressive little whistle-stop—Lamy, New Mexico. To my infinite surprise, there to meet us was Jack Calkin, a mathematician I knew well. I had met him several years before at the University of Chicago and had seen him a number of times since. Calkin had been Johnny's assistant and had gone with him to London to discuss probability problems in aerial-bombing patterns and methods. Just a few weeks before, he had joined the Manhattan Project. He was a tall, pleasant-looking man, with more savoir-faire than most mathematicians. Having heard that I was coming, he borrowed a car from the Army motor pool and drove to meet us at the train.

The sun shone brilliantly, the air was crisp and heady, and it was warm even though there was a lot of snow on the ground—a lovely contrast to the rigors of winter in Madison. Calkin drove us into Santa Fe, and we stopped for lunch at the Hotel La Fonda, where we sat at the low Spanish-style tables in the bar. After an interesting New-Mexico-style meal, we walked to a small doorway in a one-story building on a little street that bordered the central Plaza. In a modest suite of rooms, a smiling middle-aged lady invited me to fill out a few forms, turned a crank on a primitive desk machine, and produced the sheets of paper that were our passes to the Los Alamos Project. This inconspicuous little office was the entrance to the gigantic Los Alamos complex. The scene, very much like a British cloak-and-dagger mystery, brought memories of my boyhood fascination with such tales.

The project site was about forty miles northwest of Santa Fe. The ride was hair-raising, Jack having elected to show us the countryside by taking us on a short cut—a muddy track through sparse Mexican and Indian villages—until we came to the Rio Grande, which we crossed on a narrow wooden bridge.

The setting was romantic. We were going up and up into a strange, mysterious landscape of mesas, cliffs, piñon trees and brush. This became a forest of pines as we gained elevation. At a military gate in a barbed-wire fence, we showed our passes and drove on to a helter-skelter collection of one- and two-story wooden structures built along muddy, unpaved, narrow streets and paths.

We were assigned a small cottage by a pond (with the promise of larger quarters as soon as they were built). I then followed Jack to my first visit to the technical area.

We entered an office, where I was surprised to find Johnny deep in conversation with a man of middle stature, bushy eyebrows, an intense expression. He limped slightly as he paced back and forth in front of a blackboard. This was Edward Teller, to whom Johnny introduced me.

They were talking about things which I only vaguely understood. There were tremendously long formulae on the blackboard, which scared me. Seeing all these complications of analysis, I was dumbfounded, fearing I would never be able to contribute anything. However, when day after day the same equations remained and were not changed every few hours as I had expected, I regained my confidence and some hope of being able to add something to the theoretical work.

I understood snatches of their conversation, and an hour later, Johnny took me aside and explained to me formally and clearly the nature of the project and its status at the moment. The work in Los Alamos had started in earnest only about two or three months before. Von Neumann seemed very certain of its importance and radiated confidence about the ultimate success of the enterprise whose objective was the construction of an atomic bomb. He told me of all the possibilities which had been considered, of the problems relating to the assembling of fissionable materials, about plutonium (which did not yet physically exist even in the most microscopic quantities at Los Alamos). I remember very well, when a couple of months later I saw Robert Oppenheimer running excitedly down a corridor holding a small vial in his hand, with Victor Weisskopf trailing after him. He was showing some mysterious drops of something at the bottom of the vial. Doors opened, people were summoned, whispered conversations ensued, there was great excitement. The first quantity of plutonium had just arrived at the lab.

Needless to say, I soon ran across most of the Wisconsinites who had so mysteriously disappeared from Madison before us. I met Hans Bethe on the first day. I knew more about him than about Teller. I gradually met the entire group of theoretical and experimental physicists. I had known many mathematicians in Europe and in this country, but not as many physicists.

I had some knowledge of theoretical physics, despite the joke I had told Johnny about my not understanding even the autocatalytic action of a toilet. Astronomy, of course, had been my first interest, and then physics and mathematics. I had even given a course in classical mechanics at Harvard, but it is one thing to know about physics abstractly, and quite another to have a practical encounter with problems directly connected with experimental data, such as the very novel technology which was to come from Los Alamos.

I found out that the main ability to have was a visual, and also an almost tactile, way to imagine the physical situations, rather than a merely logical picture of the problems.

The feeling for problems in physics is quite different from purely theoretical mathematical thinking. It is hard to describe the kind of imagination that enables one to guess at or gauge the behavior of physical phenomena. Very few mathematicians seem to possess it to any great degree. Johnny, for example, did not have to any extent the intuitive common sense and "gut" feeling or penchant for guessing what happens in given physical situations. His memory was mainly auditory, rather than visual.

Another thing that seems necessary is the knowledge of a dozen or so physical constants, not merely of their numerical value, but a real feeling for their relative orders of magnitude and interrelations, and, so to speak, an instinctive ability to "estimate."

I knew, of course, the values of constants like the velocity of light and maybe three or four other fundamental constants—the Planck constant h, a gas constant R, etc. Very soon I discovered that if one gets a feeling for no more than a dozen other radiation and nuclear constants, one can imagine the subatomic world almost tangibly, and manipulate the picture dimensionally and qualitatively, before calculating more precise relationships.

Most of the physics at Los Alamos could be reduced to the study of assemblies of particles interacting with each other, hitting each other, scattering, sometimes giving rise to new particles. Strangely enough, the actual working problems did not involve much of the mathematical apparatus of quantum theory although it lay at the base of the phenomena, but rather dynamics of a more classical kind—kinematics, statistical mechanics, large-scale motion problems, hydrodynamics, behavior of radiation, and the like. In fact, compared to quantum theory the project work was like applied mathematics as compared with abstract mathematics. If one is good at solving differential equations or using asymptotic series, one need not necessarily know the foundations of function space language. It is needed for a more fundamental understanding, of course. In the same way, quantum theory is necessary in many instances to explain the data and to explain the values of cross sections. But it was not crucial, once one understood the ideas and then the facts of events involving neutrons reacting with other nuclei.

Teller, in whose group I was supposed to work, talked to me on that first day about a problem in mathematical physics that was part of the necessary theoretical work in preparation for developing the idea of a "super" bomb, as the proposed thermonuclear hydrogen bomb was then called. The idea of thermonuclear reactions that would release enormous amounts of energy was, of course, older. Their role in the reactions in the interior of stars was discussed in theoretical papers in the 1930s by the physicists Geoffrey S. Atkinson and Fritz Houtermans. The idea of using a uranium fission explosion to trigger a thermonuclear reaction can be credited to Teller, Bethe, Konopinski, I believe, and perhaps some others.

Teller's problem concerned the interaction of an electron gas with radiation, and it had more to do with the thermonuclear possibilities than with the assembly of the fission bomb, which was the main problem and work of Los Alamos. He guessed a formula for energy transfers connected with the so-called Compton effect about the rate of energy transfer. This formula, based on dimensional grounds and his intuition alone, was quite simple; he asked me to try to derive it more rigorously. As it was presented, there was no numerical factor in front. This seemed curious to me. I asked him explicitly about this a day or two later, and he said "Oh, the factor should be 1."

This was the first technical problem in theoretical physics I had ever tackled in my life, and I approached it from a very elementary point of view. I read papers on statistical mechanics, on properties of the radiation field, and started working with rather naive and common-sense kinematic pictures. I tried some arithmetic and obtained a formula much like Teller's, but with a numerical factor of about four in front as the rate of transfer. It was a messy little job. Edward was not satisfied with my rather elementary derivations.

Shortly after I had discussed this work with Teller, a young, more professional mathematical physicist, Henry Hurwitz, Jr., joined Teller's group and with his much better mathematical techniques and experience in the special functions that were used in this type of problem, he obtained a formula, much more scholarly than mine, involving Bessel functions. Indeed, the exact numerical factor was not very different from four. If I remember correctly, it was a root of a certain Bessel function.

The idea was to have some thermonuclear material—deuterium—next to the fission bomb, and to let it ignite after the uranium bomb had exploded. How to do it in detail was the big problem, and it was by no means easy to see how such an arrangement would ignite and not just sputter and fizzle out. There was also, theoretically at least, the hazard of getting more of an explosion than intended and of having the whole atmosphere of the earth ignite! The well-known physicist Gregory Breit was involved in calculating the chances of the ignition of the atmosphere. These, of course, had to be zero before one could even think of tampering with thermonuclear reactions on earth.

I think it was Bethe, with Emil Konopinski, a well-known theoretical physicist, who suggested tritium instead of deuterium as a material easier to ignite, given the temperature of the fission bomb. Such an engineering suggestion from theoretical work came from his superb knowledge of theoretical nuclear physics.

Bethe was the head of the theoretical division, as it was called. Actually it was his and Robert F. Bacher's papers in *Reviews of Modern Physics* which were used as the "bible" of the Los Alamos scientists, for they contained the bulk of the theoretical ideas and experimental facts known at the time. Bethe, now a Nobel Prize-winner for his earlier discovery of the mechanism of energy generation in the sun and other stars (the so-called carbon cycle), is, among other things, a virtuoso in the techniques of mathematical physics. As Feynman once put it, at Los Alamos he was, with his rigorous and definitive work, like a battleship moving steadily forward, surrounded by a flotilla of smaller vessels, the younger theoretical workers of the laboratory. He is one of the few persons for whom I merely had respect in the beginning but over the years have continuously developed liking and admiration.

When I first met Teller, he appeared youthful, always intense, visibly ambitious, and harboring a smoldering passion for achievement in physics. He was a warm person and clearly desired friendship with other physicists. Possessing a very critical mind, he also showed quickness, sense, and great determination and persistence. However, I think he also showed less feeling for true simplicity in the more fundamental levels of theoretical physics. To exaggerate a bit, I would say his talents were more in the direction of engineering, construction, and the surveying of existing methods. But undoubtedly he also had great ingenuity.

Teller was well known for his work on molecules, but he may have considered this as a sort of secondary field. I think it was the ease with which Gamow had new ideas without any technical arsenal at his disposal that pushed Teller into trying to emulate him and to attempt more fundamental work.

After he got into personal difficulties with Teller on the organizational features of the hydrogen work, Gamow later told me that before the war Teller was, in his view, a different person—helpful, willing, and able to work on other people's ideas without insisting on everything having to be his own. According to Gamow, something changed in him after he joined the Los Alamos Project.

Of course, many physicists who were almost congenitally ivory-tower types got their heads turned with the sudden realization of not only the practical but worldwide historical importance of their work—not to mention the more trivial but obvious matter of the enormous sums of money and physical facilities that surpassed anything in their previous experience. Perhaps this played a role in the personality change of some principals; with Oppenheimer, the director, it may have had a bearing on his subsequent activities, career, ideas, and role as a universal sage. Like Teller, Oppenheimer may have had a feeling of inadequacy as compared with the creators of great new physical theories. He was equal to or even more brilliant and quick than Teller, but perhaps lacked the ultimate creative spark of originality. With his fantastic intelligence, he must have realized this himself. In speed of understanding and in critical ability, he probably surpassed Bethe or Fermi.

Teller wanted to have his own stamp on much of the essential work of Los Alamos, at first via his own approach to the fission bomb. He was pushing for milder explosions, dilution of material, etc. In spite of calculations by Konopinski and others that gave a poor outlook for some of these plans, he was trying by every means to have his own adopted. Collaboration with Bethe, who was head of the theoretical division, became increasingly difficult.

As disagreements between Teller and Bethe became more frequent and acute, Teller threatened to leave. Oppenheimer, who did not want to lose such a brilliant scientist, agreed to let him and his group work in a more future-oriented field, independent of the project's main line. This is how Teller began to concentrate and organize the theoretical work for the "Super." Konopinski, Weisskopf, Serber, Richard Feynman, William Rarita, and many others all had special contributions to make, but it was really Teller who kept the thing together and moving forward during the war.

After Fermi's arrival, Teller's group became a part of Fermi's division. Fermi took great interest in the theoretical work on the thermonuclear reactions and H-bomb possibilities; at the end of the war he gave a series of lectures magnificently summarizing the work done until then—thanks mostly to the investigations of Teller and his group.

But even then, before the success of the atomic bomb itself, some of Teller's actions gave clues to what led to much of the unpleasantness and waste of time in the so-called H-Bomb controversy.

Teller's group was composed of a number of very interesting young physicists, younger even than Teller, Konopinski, or myself. It included Nick Metropolis, a Greek-American with a wonderful personality; Harold and Mary Argo, a husband-and-wife team, eager and talented; Jane Roeberg, a young woman who gave the impression of being competent; and a few others whose names I do not now remember.

There was, of course, much contact with other groups of physicists who were on the borderline of problems concerning the possibilities of a "super"; discussions with them were frequent, pleasant, and concerned many different branches of physics. One could hear about the pros and cons of the idea of implosion, which was new and vigorously debated in many of the offices. These discussions were completely open. Nothing was concealed from anybody who was a scientist.

The more formal way of letting people know what was going on was the weekly colloquia, which were held in a big hangar that also served as the movie theater. These talks covered progress of the work of the whole laboratory and the specific problems which the project encountered. They were run by Oppenheimer himself.

As for myself, after this first work on Edward's problem, I spread out my interests to other related questions, one being the problem of statistics of neutron multiplication. This was more tangible for me from the purely mathematical side. I discussed such problems of branching and multiplying patterns with David Hawkins. We wrote a report on multiplicative branching processes, which had some practical application and relevance to the problem of the initial detonation of the bomb by a few neutrons. This problem was also studied by Stan Fraenkel and by Feynman, in a more technical and classical way. Our paper could be considered the beginning of what would come to be known in mathematics as branching processes theory, a sub-field of probability theory.

I also talked a lot with von Neumann and Calkin about problems of hydrodynamics, especially those concerning the process of implosion. Somewhat to my surprise I found my purely abstract intellectual habits as a mathematician immediately useful in the work with these more practical, special, and tangible problems. I have never felt the "gap" between the mode of thinking in pure mathematics and the thinking in physics, on which many mathematicians place so much stress. Anything amenable to mental analysis was congenial for me. I do not mean the distinction between rigorous thinking and more vague "imaginings"; even in mathematics itself, all is not a question of rigor, but rather, at the start, of reasoned intuition and imagination, and, also, repeated guessing. After all, most thinking is a synthesis or juxtaposition of advances along a line of syllogisms—perhaps in a continuous and persistent "forward" movement, with searching, so to speak "sideways," in directions which are not necessarily present from the very beginning and which I describe as "sending out exploratory patrols" and trying alternative routes. It is all a multicolored thing, not very easy to describe in a way that a reader can appreciate. But I hope this kind of personal analysis of thinking in science is one of the possible interests of this book.

A discussion with von Neumann which I remember from early 1944 took several hours, and concerned ways to calculate the course of an implosion more realistically than the first attempts outlined by him and his collaborators. The hydrodynamical problem was simply stated, but very difficult to calculate—not only in detail, but even in order of magnitude.

In particular, the questions concerned values of certain numbers relating to compression versus pressure, and such. These had to be known, let us say within ten per cent or better, but the simplifications made in the outline of the calculations were of such a nature that they could not guarantee accuracy within a factor of two or three. All the ingenious shortcuts and theoretical simplifications which von Neumann and other mathematical physicists suggested, and which he tried to execute with the help of Calkin, seemed inadequate to me. In this discussion I stressed pure pragmatism and the necessity for attempting to get a heuristic survey of the general problem by simpleminded brute force—that is, more realistic, massive numerical work. At that time, in 1944, with the available computing facilities, the accuracy of the necessary numerical work could not be satisfactory. This was one of the first reasons for pressing for the development of electronic computers.

One of the charms and great attractions of life in Los Alamos in those days was the lunches at the Lodge, in the midst of friends. I was very surprised to find there and gradually to meet so many famous persons I had heard about.

Los Alamos was a very young place. At thirty-four, I was already one of the older people. What impressed me most was the very great competence of the younger people and the variety of their fields of specialization. It was almost like having an encyclopedia to look at, something that I so much like to do. I had the same feeling when talking to the young scientists around the laboratory. It is not the right expression perhaps but, roughly speaking, they were more accomplished in depth than in breadth. The older men, many of whom were European-born, had a more general knowledge. Yet science had become so ramified, specialization had proceeded so far, that it was quite difficult to retain knowledge of all the details and the overall view at the same time.

The younger scientists showed a lot of common sense in their own fields, but in general a great hesitation to engage in speculation outside their areas. Perhaps this stemmed from a fear of not being "absolutely right." Many displayed a certain anti-philosophical spirit—not anti-intellectual, but anti-philosophical. This was perhaps because of the pragmatic nature of American attitudes.

I was also struck by the well-known American talent for cooperation, the team spirit, and how it contrasted with what I had known in continental Europe. I remembered how Jules Verne had anticipated this when he wrote about the collective effort needed for the organization of his "Voyage to the Moon." People here were willing to assume minor roles for the sake of contributing to a common enterprise. This spirit of team work must have been characteristic of life in the nineteenth century and was what made the great industrial empires possible. One of its humorous side effects in Los Alamos was a fascination with organizational charts. At meetings, theoretical talks were interesting enough to the audience, but whenever an organizational chart was displayed, I could feel the whole audience come to life with pleasure at seeing something concrete and definite ("Who is responsible to whom," etc.). Organization was and perhaps still is a great American talent, although this is written at a time when the so-called energy crisis appears to me to be more a crisis of momentum than of energy (a crisis of enterprise, solidarity, common spirit, determination, and cooperation for the common good).

It is difficult to describe for the general reader the intellectual flavor, the feeling, of a scientific "atmosphere." There is no specific English word for this impression. Odor and smell have unpleasant connotations; perfume is artificial; aura is suggestive of mystery, of the supernatural. The younger scientists did not have much of an aura, they were bright young men, not geniuses. Perhaps only Feynman among the young ones had a certain aura.

Six or seven years younger than I, he was brilliant, witty, eccentric, original. I remember one day Bethe's laughter shook the corridor walls, making me rush out of my office to see what was so funny. Three doors down in Bethe's office, Feynman was standing—talking and gesticulating. He was telling the story of how he had failed his draft physical examination, re-enacting his now famous gesture: when a doctor asked him to show his hands, he chose to stretch them in front of him one with palm up, the other palm down. The doctor said, "The other side!" and he reversed both hands. This and other incidents of his physical examination had caused an explosion of laughter on the entire floor. It was on the first or second day in Los Alamos that I met Feynman, and remarked to him about my surprise that $E = mc^2$—which I of course believed in theoretically but somehow did not really "feel"—was, in fact, the basis of the whole thing and would bring about a bomb. What the whole Project was working on depended on those few little signs on paper. Einstein himself, when he was first told before the war about radioactive phenomena showing the equivalence of mass and energy, allegedly replied, "Ist das wirklich so? ist das wirklich so?" (Is that really so?)

Jokingly I told Feynman, "One day people will discover that a cubic centimeter of vacuum is really worth ten thousand dollars—it is equivalent to so much energy." He immediately agreed and added, "Yes, but of course it will have to be *pure* vacuum!" Indeed, people now know about the polarization of vacuum. The force between two electrons or two protons is not e^2/r^2, but an infinite series of which this is the first term. It works on itself, like two almost-parallel mirrors, which show a reflection of a reflection of a reflection, ad infinitum.

Writing this reminds me of a feeling I once had when I visited the cyclotron in Chicago with Fermi. He took me around and made me walk through an incredibly heavy door, which he said "would flatten you into a piece of paper were it closed on you." We walked between the poles of the magnet, and I reached into my pocket for the penknife with which I sometimes play. Suddenly it was jerked out of my hand when I touched it. The power of the vacuum! This made me physically conscious of the reality of "empty" space.

Feynman was also interested in many purely mathematical recreations not related to physics. I remember how he once gave an amusing talk about triangular numbers and managed to entertain everybody with his humor. At the same time, he was doing mathematics, and showing the foolishness of excessive cleverness and the irrationality of such strange interests.

One day he recited to me the following:
"I wonder why I wonder,
I wonder why I wonder why I wonder.
I wonder why I wonder why I wonder why,"
and so on.

It all depends on where you put the intonation, conveying a different meaning in every case. He did it marvelously in five or six styles, each with different stresses, as it were, most humorously.

Physically, Los Alamos consisted of a collection of two- and four-apartment buildings, temporary Army structures which turned out to be sturdy enough to survive for many years after the end of the war. To his everlasting credit, Oppenheimer insisted that they be laid out along the contours of the land, retaining as many trees as possible, instead of in the monotonous rectangular pattern of army camps and company towns. Still they were rather primitive, equipped with coal furnaces and coal stoves in the kitchens. People griped about the inadequacies of the housing situation, and wives had all sorts of complaints. But I found Los Alamos on the whole quite comfortable. The climate of New Mexico—Los Alamos in particular, at an elevation of seventy-two hundred feet—was one of the best I have ever lived in.

Placzek, a physicist who joined the project after the war, felt that east of the Rocky Mountains the United States was on the whole climatically uninhabitable, "unbewohnbar." This is true especially for Europeans who are not accustomed to hot and muggy summers or to penetrating winter cold. In Cambridge, I used to tell my friends that the United States was like the little child in a fairy tale, at whose birth all the good fairies came bearing gifts, and only one failed to come. It was the one bringing the climate.

Soon after my arrival in Los Alamos I met David Hawkins, a young philosopher from Berkeley, one of the people Oppenheimer had brought with him to staff the administration of the Laboratory. We hit it off intellectually right away.

Hawkins is a tallish, blue-eyed, blond descendant of

early New Mexico settlers. His father, Judge Hawkins, was a famous figure at the turn of the century. He was a lawyer and an official of the Territory, important in the Santa Fe Railroad operations. David was brought up in the small community of La Luz, in the southern part of the State. I mention this because later, when the bomb was exploded in the Jornada del Muerte desert near Alamagordo, David worried that blinding flashes or the heat and shock phenomena might be dangerous for people living in La Luz, some thirty or forty miles away, where his sister had her home.

Hawkins is a man of wide interests, with great breadth of knowledge, very good education, and a very logical mind. He regards scientific problems not as a narrow specialist, but from a general epistemological and philosophical point of view. To top it off, he is the most talented amateur mathematician I know. He told me that at Stanford he took some courses from Ouspenski, the Russian émigré specialist in probability and number theory, but he has not had any extensive training in mathematics. He has a very great natural feeling for it and a talent for manipulation. He is the most impressive of the non-professional mathematicians or physicists I have met anywhere in the world.

We discussed problems of neutron chain reactions and the probability problems of branching processes, or multiplicative processes, as we called them in 1944.

I was interested in the purely stylized problem of a branching tree of progeny from one neutron which may multiply, into zero (that is, the death of a neutron by absorption), or one (that just continues itself), or two or three or four (that is, causes the emergence of new neutrons), each possibility with a given probability. The problem is to follow the future course and the chain of possibilities through many generations.

Very early Hawkins and I detected a fundamental trick to help study such branching chains mathematically. The so-called characteristic function, a device invented by Laplace and useful for normal "addition" of random variables, turned out to be just the thing to study "multiplicative" processes. Later we found that observations to this effect had been made before us by the statistician Lotka, but the real theory of such processes, based on the operation of iteration of a function or of operators allied to the function (a more general process), was begun by us in Los Alamos, starting with a short report. This work was strongly generalized and broadened in 1947, after the war, by Everett and myself after he joined me in Los Alamos. Some time later, Eugene Wigner brought up a question of priorities. He was eager to note that we did this work quite a bit before the celebrated mathematician Andrei N. Kolmogoroff and other Russians and some Czechs had laid claim to having obtained similar results.

I liked Hawkins's general curiosity, his almost unique knowledge of the fundamentals of several scientific theories—not only in the conceptual elements of physics but in biology and even economics. I liked his interest and genuinely original work in what was to become known as "information theory" after it became formalized by Wiener and especially by Claude Shannon. David applied to economic problems the mathematical ideas of von Neumann and Morgenstern in game theory.

Hawkins has since written several interesting papers and an excellent book on the philosophy of science, or rather on the philosophy of rational thinking, called *The Language of Nature*.

Hawkins's position in Los Alamos at first was as a liaison between Oppenheimer's office and the military. Some years later he wrote two volumes, since declassified, about the organization and the scientific history from the early days of Los Alamos until the end of the war. I did not know it at the time (and it was not obvious from conversations with him) that in the nineteen thirties he had been involved on the West Coast with communist sympathizer groups. That caused him great trouble before and during the McCarthy era, including hearings in Washington. He came out of it completely vindicated.

His wife, Frances, an extremely interesting person, became friendly with Françoise, and we saw each other a great deal. At the time of my illness in California in 1946, the Hawkinses were immensely helpful to us in caring for our daughter Claire, who was then an eighteen-month-old infant.

Hawkins left Los Alamos after the end of the war to take the post of professor of philosophy at the University of Colorado in Boulder, where he is today.

The Los Alamos community was completely different from any where I had ever lived and worked. Even Lwów, which had a dense concentration of people and where the mathematicians and university people were in daily contact and spent much time together in restaurants and coffee houses, did not have the degree of togetherness of Los Alamos. It was even more pronounced there because of the isolation and the smallness of the town, and the proximity of all the buildings. People visited each other constantly at all hours after work. What was novel to me was that these were not mathematicians (except von Neumann and two or three younger persons), but physicists, chemists, and engineers— psychologically quite different from my more inward-oriented mathematical colleagues. The variety and richness of the physicists was interesting and delightful to observe. On the whole, theoreticians and experimentalists differed in temperament.

It has been said that at lunch in Fuller Lodge one could see as many as eight or ten Nobel Prize-winners eating at the same time (Rabi, Lawrence, Fermi, Bloch, Bohr, Chadwick, and others). Their interests were wide because physics has more definite and obvious central problems than mathematics, which splits into many almost independent domains of thought. They considered not only the main problem—the construction of an atomic bomb and related physical questions about phenomena that would attend the explosion—the strictly project work—but also general questions about the nature of physics, the future of physics, the impact of nuclear experiments on the technology of the future, and contrastingly its influence on the future development of theory. Beyond this, I remember very many after-dinner discussions about the philosophy of science, and of course on the world situation, from daily progress on the war fronts to prospects of victory in the months to come.

The intellectual quality of so many interesting persons and their being constantly together was unique. In the entire history of science there had never been anything even remotely approaching such a concentration. The radar project in Cambridge, Massachusetts, proceeding at the same time, had some of these characteristics, but without the same intensity. It was more technological perhaps, and did not touch as many fundamental questions of physics.

Who were some of the luminaries of this fantastic assembly? Von Neumann, Fermi, Bethe, Bohr, Feynman, Teller, Oppenheimer, O. R. Frisch, Weisskopf, Segré, and many more. I have already tried to sketch the personalities of some of them and can describe a few more.

I first met Fermi when he arrived at Los Alamos, a few months after us, after the Chicago pile had been successfully completed. I remember sitting at lunch in Fuller Lodge before his arrival with six or seven people, including von Neumann and Teller. Teller said, "It is quite certain now that Enrico will arrive next week." I had learned earlier that Fermi was referred to as "the pope" because of the infallibility of his pronouncements. So immediately I intoned: "Annuncio vobis gaudium maximum, papam habemus," which is the classical way cardinals announce the election of a pope on the balcony overlooking St. Peter's Square, after the white smoke comes out of the chimney in the Vatican. Johnny, who understood, explained this reference, and the allusion was applauded by the entire table.

Fermi was short, sturdily built, strong in arms and legs, and rather fast moving. His eyes, darting at times, would be fixed reflectively when he was considering some question. His fingers often nervously played with a pencil or a slide rule. He usually appeared in good humor, with a smile almost perpetually playing around his lips.

He would look at a questioner in an inquiring way. His conversation included many questions rather than expressions of opinion. His questions were formulated in such a way, however, that it was clear which way Fermi's beliefs or guesses went. He would try to elucidate other persons' thoughts by asking questions in a Socratic manner, yet more concretely than in Plato's succession of problems.

Sublimated common sense characterized his thoughts. He had will power and control; and not obstinacy but persistence in following a line, all the while looking very carefully at possible ramifications. He would not neglect the opportunities that presented themselves, often by chance, from random observations in scientific work.

Once when we discussed another physicist, he characterized him as too systematically obstinate. Yet he also told me that he liked to work very systematically in an orderly fashion in order to keep everything under control. At the same time he had decided in his youth to spend at least one hour a day thinking in a speculative way. I liked this paradox of a systematic way of thinking unsystematically. Fermi had a whole arsenal of mental pictures, illustrations, as it were, of important laws or effects, and he had a great mathematical technique, which he used only when necessary. Actually it was more than mere technique; it was a method for dissecting a problem and attacking each part in turn. With our limited knowledge of introspection this cannot be explained at the present time. It is still an "art" rather than a "science." I would say that Fermi was overwhelmingly rational. Let me explain what I mean: the special theory of relativity was strange, irrational, seen against the background of what was known before. There was no simple way to develop it through analogies with previous ideas. Fermi probably would not have tried to develop such a revolution.

I think he had a supreme sense of the important. He did not disdain work on the so-called smaller problems; at the same time, he kept in mind the order of importance of things in physics. This quality is more vital in physics than in mathematics, which is not so uniquely tied to "reality." Strangely enough, he started as a mathematician. Some of his first papers with very elegant results were devoted to the problem of ergodic motion. When he wanted to, he could do all kinds of mathematics. To my surprise, once on a walk he discussed a mathematical question arising from statistical mechanics which John Oxtoby and I had solved in 1941.

Fermi's will power was obvious, even to the extent of controlling his impulsive gestures. In my opinion, he deliberately avoided volatile Latin mannerisms, and perhaps by a conscious decision controlled gesticulations and avoided exclamations. But Enrico smiled and laughed very readily.

In all activities, scientific or otherwise, he had a mixture of semi-logical whimsical humor about common-sense points of view. When he played tennis, for instance, if he lost four games to six, he would say: "It does not count because the difference is less than the square root of the sum of the number of games." (This is a measure of purely random fluctuations in statistics.)

He loved political discussions, and he loved trying—not too seriously—to foresee the future. He would ask people in a group to write down what they thought would happen, and put it in a sealed envelope to be opened a couple of months later. On the whole he was very pessimistic about the long-range outlook politically, concluding that humanity is still foolish and would destroy itself one day.

He could be also quite a tease. I remember his Italian inflections when he would taunt Teller with statements like: "Edward-a how com-a the Hungarians have not-a invented anything?" Once Segré, who was very fond of fishing on weekends in the streams of the Los Alamos mountains, was expounding on the subtleties of the art, saying that it was not easy to catch trout. Enrico, who was not a fisherman, said with a smile, "Oh, I see, Emilio, it is a battle of wits."

In conversations with friends about the personalities of others, he tried to be entirely detached and objective, allowing little of personal or subjective opinions or feelings to surface. About himself, he had tremendous self-assurance. He knew that he had the touch as well as luck on top of his supreme common sense, enormous mathematical technique, and knowledge of physics.

Enrico was fond of walking; several times we walked all the way from Los Alamos down the walls of a canyon and along a stream to the Bandelier National Monument. It was a walk of seven or eight miles during which we had to cross the stream more than thirty times. The walk lasted several hours, and we discussed many subjects.

I should mention here one of my own peculiarities: I do not like walking uphill. I don't really know why. Some people tell me that I tend to go too fast from impatience and get winded for that reason. I do not mind walking on level ground and I actually enjoy walking down hill. Years ago I bought a German travel guidebook called "One Hundred Downhill Walks in the Alps." Certainly a humorous title.

After the war, on one of these downhill excursions in Frijoles Canyon, I told Fermi how in my last year of high school I was reading popular accounts of the work of Heisenberg, Schrödinger, and De Broglie on the new quantum theory. I learned that the solution of the Schrödinger equation gives levels of hydrogen atoms with a precision of six decimals. I wondered how such an artificially abstracted equation could work to better than one part in a million. A partial differential equation pulled out of thin air, it seemed to me, despite the appearances of derivation by analogies. I was relating this to Fermi, and at once he replied: "It [the Schrödinger equation] has no business being that good, you know, Stan."

He went on to say that in the fall he intended to give a really logical introduction and derivation of quantum theory in his course at the University of Chicago. He apparently worked at it, but told me the next summer, when he returned to Los Alamos, "No, I didn't succeed to my satisfaction in giving a really rational introduction to quantum theory." It is not just a question of axioms as some naive purist might think. The question is why such and no other axiom? Any working algorithm can be axiomatized. How to introduce, justify, tie up or simplify the axioms, historically or conceptually, and how to base them on experiments—that is the problem.

Von Neumann and Fermi were really quite different in personality. Johnny was perhaps broader in his interests than Enrico. He had more specifically expressed interests in other fields, certainly, for example, in ancient history. Fermi did not show any great interest in or liking for the arts. I never remember him discussing music, painting, or literature. Current affairs, politics, yes; history, no. Von Neumann was interested in both. Fermi did not indulge in quotations or allusions, Latin or otherwise, although he liked epigrammatic formulations occasionally. But he did not display a gymnasium or lycée type of education or the resultant mental habits. His overwhelming characteristic was his Latin clarity. Von Neumann did not consciously insist on simplicity; on the contrary, he liked to show clever complications on occasion.

In their lectures to students or scientific gatherings, they demonstrated their different approaches. Johnny did not mind showing off brilliancy or special ingenuity; Fermi, on the contrary, always strived for the utmost simplicity, and when he talked everything appeared in a most natural, direct, bright, clear light. After students had gone home, they were often unable to reconstruct Fermi's dazzlingly simple explanation of some phenomenon or his deceptively simple-looking idea on how to treat a physical problem mathematically. In contrast, von Neumann showed the effects of his sojourns at German universities. He was absolutely devoid of pomposity, but in his language structure he could be complicated, though perfect logic always gave a unique interpretation to his words.

They held high opinions of each other. I remember a discussion of some hydrodynamical problem Fermi had been thinking about. Von Neumann showed a way to consider it, using a formal mathematical technique. Fermi told me later with admiration, "He is really a professional, isn't he!" As for von Neumann, he always took external evidences of success seriously; he was quite impressed by Fermi's Nobel Prize. He also appreciated wistfully other people's ability to get results by intuition or seemingly pure luck, especially by the apparent effortlessness of Fermi's fundamental physics discoveries. After all, Fermi was perhaps the last all-around physicist in the sense that he knew the theory, did original work in many branches, and knew what experiments to suggest and even do himself; he was the last to be great both in theory and as an experimenter.

Niels Bohr, the discoverer of the quantized electron orbits in the atom and a great pioneer of quantum theory, was in Los Alamos for several months. He was not very old. To me at thirty-five, he seemed ancient, even though Bohr in his late fifties was very active and energetic, physically as well as mentally. He walked, skied, and hiked in the Los Alamos mountains. Somehow he seemed the embodiment of wisdom. (Wisdom, perhaps not genius in the sense of Newton or Einstein.) He knew what not to attempt and how much could be done without mathematics, which he left to others. This enormous wisdom is what I liked about him.

Departing from his usual caution about expressing opinions about other people, Fermi remarked once that when Bohr talked he sometimes gave the impression of a Catholic priest celebrating mass. It was an iconoclastic statement, since so many physicists are still under the spell of Bohr.

He had his own kind of genius that made him a great physicist but, to my mind, some of his students were almost benighted by his complementarity philosophy of "one can say this, but on the other hand one can . . ." or "one cannot say sharply what this means." People without his great sense and intuitive wisdom were led astray and lost the precision and sharpness of their intellectual or scientific approach, in my opinion. But he still has many admirers. Victor F. Weisskopf is one.

It seems to me that as a philosophical guideline, complementarity is essentially negative. It can only console. Whether it can be positively useful other than in philosophical consolations is a question which troubles me.

Bohr's speech was very difficult to understand, and anecdotes about him abound. Most of the time it was impossible to get his exact words. One day a young physicist, Ruby Scherr, was called on the public address system. Here I should explain that every day at periodic intervals the halls of the laboratory resounded with announcements and requests, the most frequent of which was a call for J. J. Gutierrez, who was a supply factotum and jack of all trades. Other calls were requests for the return of such and such an instrument, or even the Sears, Roebuck catalogue. One day among other announcements came one asking Ruby Scherr to please go to Nicholas Baker's office. (Nicholas Baker was the pseudonym of Bohr for security reasons; Fermi's was Farmer.) As Ruby Scherr tells the story, he went to the office, saw several physicists sitting around and obviously listening to a presentation by Bohr. Bohr stopped, mumbled a few incomprehensible sentences in the direction of Scherr, and suddenly ended with a crystal-clear three words: "Guess how much?" Scherr, who had not understood a word of the question, blushed with embarrassment, shook his head shyly and remained silent. After a moment Bohr again in a clear voice said "10^{41}." Whereupon everybody laughed. To this day, Scherr does not know what it was all about.

Another Bohr story illustrates the absentmindedness of scientists: it was well known throughout Los Alamos that Nicholas Baker was Niels Bohr; nevertheless, his true name was never supposed to be mentioned in public. At one colloquium, Weisskopf referred to "the well-known Bohr principle." "Oh excuse me," he fumbled, "the Nicholas Baker principle!" General laughter greeted this security breach.

Not all of us were unduly security conscious. Every scientist, old or young, had in his office a safe where secret documents had to be kept. Indeed, the Project must have had more safes than all the banks in New York. Once in Bohr's office I watched him struggle to open his safe. The safes could be opened with a rather simple combination of three two-digit numbers. He tried and tried for a long time, finally succeeding. He pulled out the drawer and exclaimed delightedly: "I believe I have done enough for the day." The story that Dick Feynman could open safes whose combinations had been forgotten by their owners is true. He apparently listened to the clicks of the tumblers and sometimes he guessed which combinations of digits of numbers like π or e in mathematics, or c, the velocity of light, or h, Planck's constant, had been selected by the owners for the combinations.

One thing that relieved the repetition and alternation of work, intellectual discussions, evening gatherings, social family visits and dinner parties, was when a group of us would play poker about once a week. The group included Metropolis, Davis, Calkin, Flanders, Langer, Long, Konopinski, von Neumann (when he was in town), Kistiakowski sometimes, Teller, and others. We played for small stakes; the naïveté of the game and the frivolous discussions laced with earthy exclamations and rough language provided a bath of refreshing foolishness from the very serious and important business that was the raison d'être of Los Alamos.

In playing such a game, unless you are vitally interested in the game itself, and not merely in its relaxing qualities, you will not do well. Von Neumann, Teller, and I would think about completely unrelated subjects during the bidding or betting; consequently, more often than not we were the losers. Metropolis once described what a triumph it was to win ten dollars from John von Neumann, author of a famous treatise on game theory. He then bought his book for five dollars and pasted the other five inside the cover as a symbol of his victory. It may not be clear to non-scientists or non-mathematicians that one can do theoretical work in one's head and pursue it quite intensely while literally carrying on some other more prosaic activity.

The Trinity test, Hiroshima, V-J Day, and the story of Los Alamos exploded over the world almost simultaneously with the A-Bomb. Publicity over the secret wartime Project filled the newspapers and its administrative heads were thrown in the limelight. In one newspaper interview out of many published the day after Hiroshima, E. O. Lawrence "modestly admitted," according to the interviewer, "that he more than anyone else was responsible for the atomic bomb." Similar statements by and about others filled the media. Oppenheimer was reported to have described his feelings after the unearthly light of the initial flash of the Trinity experiment by quoting from the Hindu epic, the Bhagavad Gita: "It flashed to my mind that I had become the Prince of Darkness, the destroyer of Universes."

What is true is that as I was reading this item in a newspaper, something else flashed through my mind, a story of a "pension" in Berlin before the war. I told it immediately to Johnny, who was eating dinner in our house. The Berlin boarders were sitting around a table for dinner and dishes were passed for each person to help himself to his share. One man was taking most of the asparagus that was on the platter. Whereupon another man stood up shyly and said: "Excuse me, Mr. Goldberg, we also like asparagus!" And the expression "asparagus" became a code word in our private conversations for trying to obtain an unduly large share of credit for scientific work or any other accomplishment of a joint or group character. Johnny loved this story so much that in our humorous conversations we played on developing the theme. We would plan to write a twenty-volume treatise on "Asparagetics through the Ages." Johnny would do "Die Asparagetics im Altertum" and I the final volume "Rückblick und Ausblick" in the manner of heavy German scholarship. Later, Carson Mark put his own stamp on these jokes by composing a song, "Oh, How I Love Asparagus," to the tune of a current popular song.

But levities like these could hardly alleviate the general feeling of foreboding upon entering into the era of history that would be called the Atomic Age. The war was over, the world and the nation had to reorganize themselves. Life would never be the same.

CHAPTER 9

Southern California
1945–1946

THE war was over and the world was emerging from the ashes. Many people left Los Alamos, either to return to their former universities like Hans Bethe, or to go to new academic positions like Weisskopf to MIT or Teller to Chicago. The government had not reached any decision yet about the fate of the wartime laboratory.

The University of Chicago took steps to start a great new center for nuclear physics, with Fermi, Teller and several others from the Manhattan District Project. Von Neumann, better than anyone else, it seems to me, argued that as a result of the role science had played in the winning of the war, the post-war academic world would not be recognizable in pre-1939 terms.

On the purely personal plane, I had no evidence that any member of my immediate family had survived (two cousins did reappear many years later, one in France, the other in Israel). Françoise had lost her mother in the concentration camp of Auschwitz. We were both American citizens now, the United States was our country, and the idea of returning to Europe never entered our heads. But the question of what job to return to from war work was very much on our minds.

I had some correspondence with Langer, who was then the chairman, about returning to Madison. He was very honest and open, and he told me with admirable frankness when I inquired about my chances for promotion and tenure: "No reason to beat around the bush, were you not a foreigner, it would be much easier and your career would develop faster." So it seemed that my chances in Wisconsin were not very good, and I looked elsewhere. Elsewhere came in the form of a letter from an old Madison friend, Donald Hyers, who had become a professor at the University of Southern California in Los Angeles. Hyers was well established there, and he asked whether I would be interested in joining the faculty as an associate professor at a salary somewhat higher than the one in Madison. The university was small, not very strong academically, and certainly not a very prestigious place, but the professors there, he said, were engaged in vigorous attempts at improving the academic standing of the institution. He invited me for a visit, and I flew to Los Angeles in August of 1945.

This was the first time I saw that city, and it gave me a very strange impression. It was a different world from any I had known, climatically, architecturally, and otherwise. I mentioned this job possibility to Johnny, and although he was rather surprised at my interest in this rather modest opportunity, he did not react negatively. His tendency was to go along. I did not see much sense in marking time in Los Alamos after the war, so I accepted the USC offer.

In early September of 1945, I went to Los Angeles to look for housing and to prepare our move from Los Alamos. In the immediate postwar period, the housing situation in Los Angeles was critical. Since we did not own a car, we were restricted to searching for a house in the vicinity of the

University. I used to say that any two points in Los Angeles were at least an hour's drive apart, a "discrete" topological space. I managed to sublet for one semester a typical small Los Angeles house on a modest street lined with spindly palm trees. To me it seemed adequate, but it appeared rather miserable to Françoise. Nevertheless, we settled there temporarily for lack of anything better. I noticed that in our various moves from one habitat to the next all our material possessions, clothes, books, furnishings had a way of diminishing in transit. I used to say that they dwindled to 1/e, in analogy to the energy losses of particles in transit through "one mean free path."

For the second semester of that academic year (1945–46), Hal and Hattie von Breton, good friends of the Hawkinses, invited us to stay in their summer cottage on Balboa Island across from Newport Beach. It was on the water, beautiful and comfortable—a wonderful change from the university neighborhood but a little too far for me to commute daily— so during the week I lived in a hotel near the campus and went home to the island on weekends. Françoise remained on Balboa with our baby daughter Claire, who had been born in Los Alamos the year before.

At USC I found the academic atmosphere somewhat restricted, rather anticlimactic after the intensity and the high level of science at Los Alamos. Everyone was full of good will, even if not terribly interested in "research." The "teaching load" to which I was reluctantly returning was not too heavy. All in all, things looked promising had it not been for a violent illness which struck me suddenly. I had returned to Los Angeles from a mathematics meeting in Chicago with a miserable cold. It was a stormy day; on the walk from the bus to the house in Balboa the violent winds almost choked me. That same night I developed a fantastic headache. Never in my life had I experienced a headache of any kind; this was a new feeling altogether—the most severe pain I had ever endured, all-pervading and connected with a sensation of numbness creeping up from the breast bone to the chin. I remembered suddenly Plato's description of Socrates after he was given the hemlock in prison; the jailor made him walk and told him that when the feeling of numbness starting in the legs reached his head he would die.

Françoise had difficulty in finding a doctor who would come to the island in the middle of the night. The one who finally came could not find anything visibly wrong and gave me a shot of morphine to alleviate the excruciating pain. The next morning I felt almost normal but with a lingering feeling of lassitude and an inability to express myself clearly, which came and went. Nevertheless, I returned to Los Angeles and gave my lectures at the university. The following night the violent headache reappeared. When I tried to telephone Françoise from my hotel room, I noticed that my speech was confused, that I was barely able to form words. I tried to talk around the expressions which would not come out and form equivalent ones, but it was mostly a meaningless mumble—a most frightening experience. Greatly alarmed by my incoherent phone call (I don't know how I managed to remember the phone number at home), Françoise called the von Bretons and asked them to send a doctor to see me. In fact, two doctors appeared. Perplexed by my symptoms that came and went, they took me to Cedars of Lebanon hospital. A severe attack of brain troubles began, which was to be one of the most shattering experiences of my life. By the way, many of the recollections of what preceded my operation are hazy. Thanks to what Françoise told me later I was able to put it together.

For several days I underwent various tests—encephalograms, spinal taps, and the like. The encephalogram was peculiar. The doctors suspected a tumor, which could be benign or malignant. Dr. Rainey, a neurosurgeon pupil of Cushing, was called in and an operation was planned for the following day. Of all this I knew nothing, of course. I remember only trying to distract the nurse's attention by telling her to look out of the window so I could read my chart. I saw there some alarming notation about C-3 which I suspected to mean the third convolution of the brain. Through all this I was overcome by an intense fear and began to think I was going to die. I considered my chances of surviving to be less than half. The aphasia was still present; much of the time when I tried to speak I uttered meaningless noises. I do not know why no one thought of ascertaining whether I could write instead of speak.

Françoise, alerted by the von Bretons, rushed all the way back from Balboa by taxi and arrived on the scene just as I was beginning to vomit bile, turning green and losing consciousness. She feared I was dying and made a frantic telephone call to the surgeon, who decided the operation should be performed immediately. This probably saved my life; the emergency operation relieved the severe pressure on my brain which was causing all the trouble. I remember that in my semi-conscious state my head was being shaved by a barber (he happened to be a Pole) who said a few words in Polish, to which I tried to reply. I remember also returning to consciousness briefly in a pre-operating room and wondering whether I was already in the morgue. I also remember hearing the noise of a drill. This was a true sensation as it turned out, for the doctors drilled a hole in my skull to take some last-minute X-rays. The surgeon performed a trepanation not knowing exactly where or what to look for. He did not find a tumor, but did find an acute state of inflammation of the brain. He told Françoise that my brain was bright pink instead of the usual gray. These were the early days of penicillin, which they applied liberally. A "window" was left on the brain to relieve the pressure which was causing the alarming symptoms.

I remained in a post-operative coma for several days. When I finally woke up, I felt not only better, but positively euphoric. The doctors pronounced me saved, even though they told Françoise to observe me for any signs of changes of personality or recurrence of the troubles which would have spelled brain damage or the presence of a hidden growth. I underwent more tests and examinations, and the illness was tentatively diagnosed as a kind of virus encephalitis. But the disquietude about the state of my mental faculties remained with me for a long time, even though I recovered speech completely.

One morning the surgeon asked me what 13 plus 8 were. The fact that he asked such a question embarrassed me so much that I just shook my head. Then he asked what the square root of twenty was, and I replied: about 4.4. He kept silent, then I asked, "Isn't it?" I remember Dr. Rainey laughing, visibly relieved, and saying, "I don't know." Another time I was feeling my heavily bandaged head, and the doctor chided me saying the bacteria could infect the incision. I showed him I was touching a different place. Then I remembered the notion of a mean free path of neutrons and asked him if he knew what the mean free path of bacteria was. Instead of answering, he told me an unprintable joke about a man sitting on a country toilet and how the bacteria leaped from the splashing water. The nurses seemed to like me and offered all kinds of massages and back rubs and special diets, which helped my morale more than my physical condition (which was surprisingly good).

Many friends came to visit me. Jack Calkin, who was on leave on Catalina Island, appeared several times at the hospital. So did colleagues from the University. I remember the mathematician Aristotle Dimitrios Michael. He talked so agitatedly that I fell out of bed listening to him. This scared him very much. But I managed to scramble back even though I was still slightly numb on one side. Nick Metropolis came all the way from Los Alamos. His visit cheered me greatly. I found out that the security people in Los Alamos had been worried that in my unconscious or semi-conscious states I might have revealed some atomic secrets. There was also some question as to whether this illness (which was never properly diagnosed) might have been caused by atomic radiation. But in my case this was highly improbable, for I had never been close to radioactive material, having worked only with pencil and paper. University officials visited me, too. They seemed concerned about my ability to resume my teaching duties after I got well. People were acutely concerned about my mental faculties, wondering whether they would return in full. I worried myself a good deal about that, too; would my ability to think return in its entirety or would this illness leave me mentally impaired? Obviously in my profession, complete restoration of memory was of paramount importance. I was quite frightened, but in my self-analysis I noticed that I could imagine even greater states of panic. Logical thought processes are very much disturbed by fright. Perhaps it is nature's way of blocking the process in times of danger to allow instinct to take over. But it seems to me that mere instincts, which reside in nerves and in muscle "programming," are no longer sufficient to cope with the complicated situations facing modern man; some sort of reasoning ability is still needed in the face of most dangerous situations.

I regained my strength and faculties gradually and was allowed to leave the hospital after a few weeks. I obtained a leave of absence from the university.

I remember being discharged from the hospital. As I was preparing to leave, fully dressed for the first time, standing in the corridor with Françoise, Erdös appeared at the end of the hall. He did not expect to see me up, and he exclaimed: "Stan, I am so glad to see you are alive. I thought you were going to die and that I would have to write your obituary and our joint papers." I was very flattered by his pleasure at seeing me alive, but also very frightened to realize that my friends had been on the brink of giving me up for dead.

Erdös had a suitcase with him and was just leaving after a visit to Southern California. He had no immediate commitments ahead and said, "You are going home? Good, I can go with you." So we invited him to come with us to Balboa and stay awhile. The prospect of his company delighted me. Françoise was somewhat more dubious, fearing that it would tire me too much during the early part of my convalescence.

A mathematical colleague from USC drove us all back to the von Bretons' house on Balboa Island. Physically, I was still very weak and my head had not yet healed. I was wearing a skullcap to protect the incision until my hair grew back, I remember having difficulty walking around the block the first few days, but gradually my strength returned, and soon I was walking a mile each day on the beach.

In the car on the way home from the hospital, Erdös plunged immediately into a mathematical conversation. I made some remarks, he asked me about some problem, I made a comment, and he said: "Stan, you are just like before." These were reassuring words, for I was still examining my own mind trying to find out what I might have lost from my memory. Paradoxically, one can perhaps realize what topics one has forgotten. No sooner had we arrived than Erdös proposed a game of chess. Again I had mixed feelings: on one hand I wanted to try; on the other, I was afraid to in case I had forgotten the rules of the game and the moves of the pieces. We sat down to play. I had played a lot of chess in Poland and had more practice than he had, and I managed to win the game. But the feeling of elation that followed was immediately tempered by the thought that perhaps Paul had let me win on purpose. He proposed a second game. I agreed, although I felt tired, and won again. Whereupon it was Erdös who said, "Let us stop, I am tired." I realized from the way he said it that he had played in earnest.

In the days that followed we had more and more mathematical discussions and longer and longer walks on the beach. Once he stopped to caress a sweet little child and said in his special language: "Look, Stan! What a nice epsilon." A very beautiful young woman, obviously the child's mother, sat nearby, so I replied, "But look at the capital epsilon." This made him blush with embarrassment. In those days he was very fond of using expressions like *SF* (supreme fascist) for God, *Joe* (Stalin) for Russia, *Sam* (Uncle Sam) for the United States. These were for him objects of occasional scorn.

Gradually my self-confidence returned, but every time a new situation occurred in which I could test my returning powers of thought, I was beset by doubts and worries. For example, I received a letter from the Mathematical Society asking me if I would write for the *Bulletin* an obituary article on Banach, who had died in the fall of 1945. This again gave me reason to ponder. It seemed a little macabre after having barely escaped death myself to write about another's demise. But I did it from memory, not having a library around, and sent in my article with apprehension, wondering if what I had written was weak or even nonsensical. The editors replied that the article would appear in the next issue. Yet my satisfaction and relief were again followed by doubt for I knew that all kinds of articles were printed, and I did not have such a high opinion of many of them. I still felt unsure that my thinking process was unimpaired.

Normally primitive or "elementary" thoughts are reactions to or consequences of external stimuli. But when one starts thinking about thinking in a sequence, I believe the brain plays a game—some parts providing the stimuli, the others the reactions, and so on. It is really a multi-person game, but consciously the appearance is of a one-dimensional, purely temporal sequence. One is only consciously aware of something in the brain which acts as a summarizer or totalizer of the process going on and that probably consists of many parts acting simultaneously on each other. Clearly only the one-dimensional chain of syllogisms which constitutes thinking can be communicated verbally or written down. Poincaré (and later Polya) tried to analyze the thought process. When I remember a mathematical proof, it seems to me that I remember only salient points, markers, as it were, of pleasure or difficulty. What is easy is easily passed over because it can be reconstituted logically with ease. If, on the other hand, I want to do something new or original, then it is no longer a question of syllogism chains. When I was a boy I felt that the role of rhyme in poetry was to compel one to find the unobvious because of the necessity of finding a word which rhymes. This forces novel associations and almost guarantees deviations from routine chains or trains of thought. It becomes paradoxically a sort of automatic mechanism of originality. I am pretty sure this "habit" of originality exists in mathematical research, and I can point to those who have it. This process of creation is, of course, not understood nor described well enough at present. What people think of as inspiration or illumination is really the result of much subconscious work and association through channels in the brain of which one is not aware at all.

It seems to me that good memory—at least for mathematicians and physicists—forms a large part of their talent. And what we call talent or perhaps genius itself depends to a large extent on the ability to use one's memory properly to find the analogies, past, present and future, which, as Banach said, are essential to the development of new ideas.

I continue to speculate on the nature of memory and how it is built and organized. Although one does not know much at present about its physiological or anatomical basis, what gives a partial hint is how one tries to remember things which one has temporarily forgotten. There are several theories about the physical aspects of memory. Some neurologists or biologists say that it consists perhaps of permanently renewed currents in the brain, much as the first computer memories were built with sound waves in a mercury tank. Others say that it resides in chemical changes of RNA molecules. But whatever its mechanism, an important thing is to understand the access to our memory.

Experiments seem to indicate that the memory is complete in the sense that everything we experience or think about is stored. It is only the conscious access to it that is partial and varies from person to person. Some experiments have shown that by touching a certain spot in the brain a subject will seem to recall or even "feel" a situation that happened in the past—such as being at a concert and actually hearing a certain melody.

How is memory gradually built up during one's conscious or even unconscious life and thought? My guess is that everything we experience is classified and registered on very many parallel channels in different locations, much as the visual impressions that are the result of many impulses on different cones and rods. All these pictures are transmitted together with connected impressions from other senses. Each such group is stored independently, probably in a great number of places under headings relevant to the various categories, so that in the visual brain there is a picture, and together with the picture something about the time, or the source, or the word, or the sound, in a branching tree which must have additionally a number of connecting loops. Otherwise one could not consciously try and sometimes succeed in remembering a forgotten name. In a computing machine, once the address of the position of an item in the memory is lost, there is no way to get at it. The fact that we succeed, at least on occasion, means that at least one member of the "search party" has hit a place where an element of the group is stored. Thus it is common to recall a last name once the first name has been recalled.

Then I thought, how about smell? Smell is something we sense; it is not related to any sound or picture. We do not know how to call it. It has no visual impact either. Does this contradict my guesses about simultaneous storage and connections? Then I remembered the famous incident related by Proust of the smell and taste of the "madeleine" (little cake). There are many descriptions in the literature of cases where a smell previously experienced and felt suddenly brings back a long-forgotten occasion when it was first associated with a place, or a person, many years before. So, perhaps on the contrary, this is another indication.

This feeling of analogy or association is necessary to place the set of impressions correctly on the suitable end points of a sequence of branches of a tree. And perhaps this is how people differ from each other in their memories. In some, more of these analogies are felt, stored, and better connected. Such analogies can be of an extremely abstract nature. I can conceive that a concrete picture, a visual sequence of dots and dashes, may bring back an abstract thought, which apparently in a mysterious coding had something in common with it. Some part of what is called mathematical talent may depend on the ability to see such analogies.

It is said that seventy-five percent of us have a dominant visual memory, twenty-five percent an auditory one. As for me, mine is quite visual. When I think about mathematical ideas, I see the abstract notions in symbolic pictures. They are visual assemblages, for example, a schematized picture of actual sets of points on a plane. In reading a statement like "an infinity of spheres or an infinity of sets," I imagine a picture with such almost real objects, getting smaller, vanishing on some horizon.

It is possible that human thought codes things not in terms of words or syllogisms or signs, for most people think pictorially, not verbally. There is a way of writing abstract ideas in a kind of shorthand which is almost orthogonal to the usual ways in which we communicate with each other by means of the spoken or written word. One may call this a "visual algorithm."

The process of logic itself working internally in the brain may be more analogous to a succession of operations with symbolic pictures, a sort of abstract analogue of the Chinese alphabet or some Mayan description of events—except that the elements are not merely words but more like sentences or whole stories with linkages between them forming a sort of meta- or super-logic with its own rules.

For me, some of the most interesting passages about the connections between the problem of time, as involved in the memory, and the physical or even mathematical meaning of it, whether it is classical or relativistic, were written, not by a physicist or a neurologist or a professional psychologist, but by Vladimir Nabokov in his book *Ada*. Some utterances by Einstein himself, as quoted in his biographies, show the great physicist's wonder at what living in time means, since we experience only the present. But, in reality, we consist of permanent and immutable world lines in four dimensions.

With such thoughts and worries about the thinking process, I was recovering my physical strength during this period of convalescence. What comforted me the most was the receipt of an invitation to attend a secret conference in Los Alamos in late April. This became for me a true sign of confidence in my mental recovery. I could not be told on the telephone or by letter what the conference was about. Secrecy was most intense at that time, but I guessed correctly that it would be devoted to the problems of thermonuclear bombs.

The conference lasted several days. Many friends were present. Some had been directly involved, like Fraenkel, Metropolis, Teller, and myself; others were consultants, like von Neumann. Fermi was absent. The discussions were active and inquisitive. They began with a presentation by Fraenkel of some calculations on the work initiated by Teller during the war. They were not detailed or complete enough and required work on computers (not the MANIACs but other machines in operation at the Aberdeen Proving Grounds). These were the first problems attacked that way.

The promising features of the plan were noticed and to some extent confirmed, but there remained great questions about the initiation of the process and, once initiated, about its successful continuation.

(All this was to have great importance in a later lawsuit between Sperry Rand and Honeywell over the validity of patents involving computers. The claim was that computers were already in the public domain then because the government of the United States used them and therefore the patents granted later were invalid. I was one of many who were called to testify on this in 1971.)

I participated in all the Los Alamos meetings. They lasted for hours, mornings and afternoons, and I noticed with pleasure that I was not unduly tired.

I remember telling Johnny about my illness. "I was given up for dead," I said, "and thought myself that I was already dead, except for a set of measure zero." This purely mathematical joke amused him, he laughed and asked, "What measure?"

Edward Teller and Johnny were often together, and I joined them in private talks.

In one conversation they discussed the possibility of influencing the weather. They had in mind global changes, while I proposed more local interventions. For example, I remember asking Johnny whether hurricanes could not be diverted, attenuated, or dispersed with nuclear explosions. I wasn't thinking of a point source, which is symmetrical, but several explosions in a line. I reasoned that the violence and enormous energy of a hurricane lies on top of a mass of air (the weather) which itself moves gently and slowly. I wondered if one could not, even ever so slightly, change its course in time and in trajectory on the slow-moving overall weather, thus making it avoid populated areas. There are, of course, many questions and objections about such an undertaking. One of the necessary conditions would be to make detailed computations on the course of the motion of the air masses, calculations which do not exist even now. Through the years Johnny and I occasionally talked about this with experts in hydrodynamics and meteorology.

The conference over, I returned to Los Angeles. Upon alighting from the plane, two FBI agents approached me, showed their identification and asked for permission to search my luggage. A copy of the very secret Metropolis and Fraenkel report was missing, and they wondered if I might have taken it by mistake. We searched, but I did not have it. Later I learned that everybody who had attended the conference had been contacted. The authorities were very nervous, for this was potentially of grave consequence. The missing document reappeared much later among some of Teller's papers in a Los Alamos safe.

The time was rapidly approaching when I could resume teaching, but I was developing strongly negative feelings about Los Angeles. Rides through the streets where I had been driven in an ambulance reminded me of my recent illness. My feelings toward the University were colored by this, as well, and I was dissatisfied. I felt impatiently that it was not changing quickly enough from a glorified high school into a genuine institution of higher learning. I had disagreements with a dean about building up the academic level and increasing the staff. I was told he joked that he almost had a heart attack every time he saw me, even from a distance, so afraid was he that I was bringing him new proposals for expansion!

The best part of the University was the Hancock Library. It had an impressive building and some good books—but the building was better than the collection inside. The University had just acquired an old municipal library from Boston, and when I learned what it contained, I compared it to a priceless collection of hundred-year-old Sears Roebuck catalogues. This sarcastic remark probably did not enhance my popularity.

Even though I had friends like Donald Hyers, and some new acquaintances among mathematicians, physicists, and chemists, with this growing disenchantment I wanted to leave. The Los Angeles experience had not been satisfactory.

Just then I received a telegram inviting me to return to Los Alamos in a better position and at a higher salary. It was signed by Bob Richtmyer and Nick Metropolis. Richtmyer had become head of the theoretical division.

This offer to return to Los Alamos to work among physicists and live once again in the exhilarating climate of New Mexico was a great relief for me. I replied immediately that I was interested in principle. When the telegram arrived at the laboratory, it read that I was interested "in principal."

CHAPTER 10

Back at Los Alamos

1946–1949

L OS ALAMOS was at about the lowest point in its existence. Yet on returning I found that there were a number of people who had decided to stay on and that the government wanted to keep the laboratory going and have it flourish. The laboratory was to continue studies and the development of atomic bombs.

After the war there was, of course, the question of possible new wars and the weaponry of the future. I was in favor of continuing strong armament policies if only not to run the risk of being overtaken by other nations. Johnny and others were apprehensive about Russia's ability to obtain or to develop nuclear bombs, and about its intentions towards Western Europe. He was quite hawkish at that time (the words "hawks" and "doves" were not yet in use). He thought along the old historical lines of rivalries, power struggles, coalitions; he was for a Pax Americana more than some of our other physicist friends. He also foresaw early that the essential military problems would shift from the bombs themselves and their sizes and shapes to ways to deliver them— that is to say, to rocketry.

My own position was sort of halfway between him and the physicists who hoped to internationalize nuclear weapons. I thought it was naive to expect that the wolves would lie down with the lambs and felt that meaningful international agreements would take many years. One could not hope for an instant change in attitudes or in human nature itself. I distrusted the idea of the Atlantic Union as then proposed, feeling that some of the propaganda for it was too transparent. The hegemony disguised thinly under a general organization would merely raise fears and new hysterical reactions from the other side. However, I failed to realize fully the immense importance of nuclear armament and the influence it would have on the course of world events. One bomb, I told myself, was equal to a thousand-plane raid. Yet I did not realize that the power of each such bomb could be still vastly increased, and that it was possible to manufacture thousands of them. This realization came later. I felt no qualms about returning to the laboratory to contribute to further studies of the development of atomic bombs. I would describe myself as having taken a middle course between completely naive idealism and extreme jingoism. I followed my instincts (or perhaps lack of instincts) and was mainly interested in the scientific aspects of the work. The problems of nuclear physics were very interesting and led into new regions of physics and astrophysics. Perhaps I also felt that technological sequels to scientific discoveries were inevitable. Finally I trusted the ultimate good sense of humanity. The Atomic Energy Act, as finally adopted, was much more satisfactory than the initial proposals which would have left the developments of atomic energy under the sole and complete control of the military. Françoise felt more dubious morally on instinctive and emotional grounds. I always felt that it was unwise for the scientists to turn away from problems of technology. This could leave it in the hands of dangerous and fanatical reactionaries. On the other hand, the idea of merely multiplying the number of bombs to infinity made no sense whatsoever since a small fraction of the stockpile would be sufficient to destroy all population centers on the globe even if it was assumed that a majority of the missiles failed to reach their targets. I also did not believe that Russia would invade Western Europe. This was one of the supposed reasons for super-rearmament. From the Russian point of view, it seemed to me there was no possible advantage. Seeing that even in Poland, the Russians had trouble maintaining the regime, I could not see any gain for them in making West Germany communistic. On the contrary, if all of Germany were reunited under communism, it would have presented a tremendous threat to Russia. A united communist Germany would inevitably have tried to become "boss" of the communist world.

Upon our return to Los Alamos, we were given a different wartime apartment and remained in it only a few months. Jack Calkin, who was still there, lived across the street. The Hawkinses were also quite near, but were preparing to leave soon. Our resumption of a more natural, if more Spartan mode of life was a refreshing change from the rather artificial Los Angeles atmosphere.

As it turned out, I had not quite recovered all my strength from the severe illness. During the first few weeks I was back, I became tired after working only two or three hours in my office. Fortunately, this disappeared gradually, and I began to feel normal again. Apart from everything else, this illness had been a financial catastrophe. It had left me with a debt of about five thousand dollars in spite of health insurance. When, in Los Angeles, it appeared that I might die or remain permanently disabled or diminished, several of our Los Alamos friends and even persons who were merely acquaintances lent money to Françoise. This touched us very deeply. I repaid them as fast as I could. The rest took several years.

At that time, Adam, my brother, was brilliantly concluding his studies at Harvard, and came to visit us in Los Alamos. I, who had been conditioned by the prewar scarcity of jobs, was pessimistic about his chances of finding one. When I asked him what his plans were, he answered, "I'll get an instructorship, of course." I felt dubious. He must have read the skepticism in the expression on my face, because I saw in his eyes that he took me for a pessimistic old dodo. He was right, because he immediately obtained an instructor's position at Harvard and has remained there ever since. He became an eminent professor of government, and is now director of the Center for Russian Studies. He is also a prolific and successful author of books on the history of communism. Among his best known are biographies of Lenin and Stalin.

Early in 1948 we had the opportunity to move to a wing of a house on "bathtub row" which had become vacant. We remained there until we left Los Alamos twenty years later. This was a group of some five or six houses which dated from the Los Alamos Ranch School days. They were the only houses with bathtubs—all the other structures had showers. During the war, these prize dwellings had been reserved for the director and other dignitaries. Fermi, Bethe, Weisskopf, and other important scientists had lived in the modest temporary wartime constructions.

This house was located directly across from the remodeled Lodge which served as the town's hotel for VIPs and official visitors. We benefited enormously from this proximity. All our friends and acquaintances visiting Los Alamos were only a few steps away. It was easy for them to visit for a drink, a casual meal, or an hour spent on our terrace. Françoise called our house the Lodge annex. On their frequent visits to Los Alamos, Johnny and Klari particularly liked living in a little cottage just next door to our yard. The informality of all these get-togethers contributed a great deal to the pleasantness of our life in Los Alamos. No end of scientific, political and personal conversations took place there. A whole book could be written about them.

Bob Richtmyer had replaced Bethe as leader of the theoretical division. He took the place of Placzek, who had held the post for a few months after the war's end. I had met Richtmyer during the war when he visited Los Alamos periodically from Washington where he worked in the patent office. He was tall, slim, intense, very friendly, and obviously a man of great general intelligence. He was interested in many areas of mathematics and mathematical physics. Later we learned about his intense musical interests, his great linguistic talents, and his specialized skills. For example, he was very good in cryptography. But he was so extremely reserved that I found it difficult to know him intimately even though we were and are on very friendly terms.

Norris Bradbury had replaced Oppenheimer as director of the laboratory. I had met him only briefly during the war. He was a pleasant, straightforward, matter-of-fact younger man, eager to take on the responsibility for continuing this extremely important work, even though he realized that it was not easy to step into the shoes of Robert Oppenheimer who was in the process of becoming a legendary figure.

Norris deserves all the credit for rescuing the project from a slow decline into a mere "ballistics" lab. It could easily have shrunk into the narrow confines of a weapons arsenal, not unlike some that remained in the California desert. Under his management the intellectual and technological level of the laboratory began to pick up slowly but surely. It became a solid and permanent place staffed by good scientists, with an increasingly broader range of interesting scientific problems and the fantastic prospects of atomic-age technology. (Now, under Harold Agnew's direction, even more so.)

Norris was rather diffident in his approach to the scientists who had left. He felt that they should recognize by themselves how important for the country and the world it was for them to come back. As a result, although he wanted to, he did not like to ask people like Fermi or Bethe or Teller to visit. It was actually left to me, with his consent, to write such invitations, along with Carson Mark and Richtmyer. Thus, in a way I was instrumental in bringing Teller back to Los Alamos.

The laboratory began to expand again. The realization of the political importance of nuclear energy for peace and of nuclear weapons for defense made it again a most prominent vital spot in national affairs. High government officials were again frequent visitors. Jim Fisk, a former Junior Fellow at Harvard and friend of mine who had become involved in atomic energy activities and was high up in the Bell Telephone Research Labs, was one of them.

During the von Neumanns' visits, we made excursions to Santa Fe and surrounding spots, frequently eating in the small local Spanish-American restaurants.

On the road to Santa Fe, each time we drove by a place called Totavi (really more a name than a place), I would launch into Latin and recite, "Toto, totare, totavi, totatum," and he would add some form of the future. This was one of our nonsensical verbal games. Another childish one was to read road signs backwards. Johnny always read "pots" for "stop" or "otla" for "alto" in Mexico.

Another game Johnny and Klari liked to play on that road was the Black Mesa game. Black Mesa was an Indian landmark in the Rio Grande Valley which was visible on and off on the way down from Los Alamos. The first one who spotted it called the other's attention by exclaiming "Black Mesa!" and scored a point. The game went on from journey to journey, points being scored as in tennis with games and sets. They never seemed to forget from one trip to the next what the score was. Johnny always liked these brief verbal distractions from serious concentrated thought.

In the early years after the war, the AEC started to build an elegant, permanent structure for its offices and those of the security services, even before new and more comfortable housing was ready for the residents. Johnny remarked that this was entirely in the tradition of all government administrations through the ages, and he decided to call the building "El Palacio de Securita." This was a good enough mixture of Spanish, Latin and Italian. So to go him one better, I immediately named a newly built church "San Giovanni delle Bombe."

This is about the time we made up a "Nebech index." Johnny had told me the classic story of the little boy who came home from school in pre-World-War-I Budapest and told his father that he had failed his final examination. The father asked him, "Why? What happened?" The boy replied, "We had to write an essay. The teacher gave us a theme: the past, the present, and the future of the Austro-Hungarian Empire." The father asked, "So, what did you write?" and the boy answered, "I wrote: Nebech, nebech, nebech." "That is correct," his father said. "Why did you receive an F?" "I spelled nebech with two bb's," was the answer.

This gave me the idea of defining the nebech index of a sentence as the number of times the word nebech could be inserted in it and still be appropriate, though giving a different flavor to the meaning of the sentence according to the word it qualifies. For instance, one could argue that the most perfect "nebech three" sentence is Descartes' statement: Cogito, ergo sum. One can say, Cogito nebech, ergo sum. Or Cogito, ergo nebech sum. Or Cogito, ergo sum nebech. Unfortunately this elegant example occurred to me only after Johnny's death. Johnny and I used this index frequently during mathematical talks, physics meetings or political discussions. We would nudge one another, whisper "Nebech two" at a particular statement, and enjoy this greatly.

Now, if the reader is sufficiently mystified, I will explain that "nebech" is an untranslatable Yiddish expression, a combination of commiseration, scorn, drama, ridicule.

To try to give the flavor of the word, imagine the William Tell story as acted out in a Jewish school. In the scene where William Tell waits in hiding to shoot Gessler, an actor says, in Yiddish: "Through this street the Nebech must come." It is obvious that Gessler is a Nebech since he will be the victim of William Tell. But if nebech had been in front of the word street, then the accent would be on street, indicating that it was not much of a street. To appreciate this may take years of apprenticeship.

Some months after I returned to Los Alamos, I invited my old friend and collaborator from Madison, C. J. Everett, to join me in the laboratory. He had remained in Madison through the war; I knew from our correspondence that he was getting tired of teaching, so I proposed to him to visit and renew our collaboration. He was the first and only person ever to arrive in Los Alamos by bus for an official interview. The project always paid for a roomette on the train, or plane fare, and his modesty caused a sensation. Shortly after the interview, he moved to Los Alamos with his wife and son and there began the continuation of our collaboration in probability theory and other mathematics, then our joint work on the H-bomb.

In Madison he was already a shy and retiring man, but as time passed he became more and more of a recluse. In the early days of his stay in Los Alamos, although he was reluctant to mingle with people, he could still be coaxed into coming to our house if one made the solemn promise that no one else would be there at the same time. Later he even refused to do that, and now the only place one can see him is in his little windowless cubicle of an office or in a carrel in the excellent laboratory library.

One of the laboratory routines was the preparation of a monthly progress report. Every staff member had to turn in a brief résumé of his work and research activity. I have already said that Everett had a very excellent sense of humor, and one month when we had been exceedingly busy with our own work he turned in a report which said only, "Great progress was made on last month's progress report."

Two seminar talks I gave shortly after my return turned out to have good or lucky ideas and led to successful further developments. One was on what was later called the Monte Carlo method, and the other was about some new possible methods of hydrodynamical calculations. Both talks laid the groundwork for very substantial activity in the applications of probability theory and in the mechanics of continua.

The hydrodynamical calculations were for problems in which there was no hope for closed formulae or explicit classical analysis solutions. They could be described as a sort of "brute force" calculations using fictitious "particles" that were really not the fluid elements but abstract points. Instead of considering individual material points of the fluid, it was a matter of using the coefficients of an infinite series into which the continuum was developed as abstract particles for a global description of the fluid. The whole motion is described by some infinite series whose terms are successively less important. Considering only the first few of them, one changed the partial differential equations of several variables (or the integral equations in several variables) into ordinary or totally different equations for a finite number of abstract "particles." Some years later, the work of Francis Harlow in Los Angeles deepened, enlarged, and multiplied the scope of this approach to the calculations of motions of fluids or of compressible gases. These are now widely used. The possibilities of such methods have not yet been exhausted; they could play a great role in the calculations of air movements, weather prediction, astrophysical problems, problems of plasma physics, and others.

The second talk was on probabilistic calculations for a class of physical problems. The idea for what was later called the Monte Carlo method occurred to me when I was playing solitaire during my illness. I noticed that it may be much more practical to get an idea of the probability of the successful outcome of a solitaire game (like Canfield or some other where the skill of the player is not important) by laying down the cards, or experimenting with the process and merely noticing what proportion comes out successfully, rather than to try to compute all the combinatorial possibilities which are an exponentially increasing number so great that, except in very elementary cases, there is no way to estimate it. This is intellectually surprising, and if not exactly humiliating, it gives one a feeling of modesty about the limits of rational or traditional thinking. In a sufficiently complicated problem, actual sampling is better than an examination of all the chains of possibilities.

It occurred to me then that this could be equally true of all processes involving branching of events, as in the production and further multiplication of neutrons in some kind of material containing uranium or other fissile elements. At each stage of the process, there are many possibilities determining the fate of the neutron. It can scatter at one angle, change its velocity, be absorbed, or produce more neutrons by a fission of the target nucleus, and so on. The elementary probabilities for each of these possibilities are individually known, to some extent, from the knowledge of the cross sections. But the problem is to know what a succession and branching of perhaps hundreds of thousands or millions will do. One can write differential equations or integral differential equations for the "expected values," but to *solve* them or even to get an approximative idea of the properties of the solution, is an entirely different matter.

The idea was to try out thousands of such possibilities and, at each stage, to select by chance, by means of a "random number" with suitable probability, the fate or kind of event, to follow it in a line, so to speak, instead of considering all branches. After examining the possible histories of only a few thousand, one will have a good sample and an approximate answer to the problem. All one needed was to have the means of producing such sample histories. It so happened that computing machines were coming into existence, and here was something suitable for machine calculation.

Computing machines came about through the confluence of scientific and technological developments. On one side was the work in mathematical logic, in the foundations of mathematics, in the detailed study of formal systems, in which von Neumann played such an important role; on the other was the rapid progress of technological discoveries in electronics which made it possible to construct electronic computers. They, in turn, provided such a quantitative increase in the speed of operation so much greater than the mechanical relay machines that it produced a qualitative change and vastly improved and enlarged the use of the tool. The results are now known to everyone: computers introduced a new age in heuristic research, in communication, and in making the space age possible.

The number of applications in exact science, in the natural sciences, and in everyday life is so great that one can talk of "the age of computers and automata" as having begun.

At that time the computers were merely in *statu nascendi*. As a joke I proposed to make Monte Carlo calculations by hiring several hundred Chinese from Taiwan, gather them on a boat, have each one sit with an abacus, or even just pencil and paper, and make them produce the random numbers by some actual physical process like throwing dice. Then someone would collect the results, and total the statistics into single answers.

Von Neumann played a leading role in the launching of electronic computers. His unique combination of gifts, his interests, and traits of character suited him for that role. I am thinking of his ability, and inclination to go through all the tedious details of program planning, of executing the minutiae of putting very large problems in a form treatable by a computer. It was his feeling for and knowledge of the details of mathematical logic systems and the theoretical structure of formal systems that enabled him to conceive of flexible programming. This was his great achievement. By suitable flow diagramming and programming, an enormous variety of problems became calculable on one machine with all connections fixed. Before his invention one had to pull out wires and reconnect plug boards each time a problem was changed.

The Monte Carlo method came into concrete form with its attendant rudiments of a theory after I proposed the possibilities of such probabilistic schemes to Johnny in 1946 during one of our conversations. It was an especially long discussion in a government car while we were driving from Los Alamos to Lamy. We talked throughout the trip, and I remember to this day what I said at various turns in the road or near certain rocks. (I mention this because it illustrates what may be multiple storing in the memory in the brain, just as one often remembers the place on the page where certain passages have been read, whether it is on the left- or right-hand page, up or down, and so on.) After this conversation we developed together the mathematics of the method. It seems to me that the name Monte Carlo contributed very much to the popularization of this procedure. It was named Monte Carlo because of the element of chance, the production of random numbers with which to play the suitable games.

Johnny saw at once its great scope even though in the first hour of our discussion he evinced a certain skepticism. But when I became more persuasive, quoting statistical estimates of how many computations were needed to obtain rough results with this or that probability, he agreed, eventually becoming quite inventive in finding marvelous technical tricks to facilitate or speed up these techniques.

The one thing about Monte Carlo is that it never gives an exact answer; rather its conclusions indicate that the answer is so and so, within such and such an error, with such and such probability—that is, with probability differing from one by such and such a small amount. In other words, it provides an estimate of the value of the numbers sought in a given problem.

I gave a lot of "propaganda" talks for this method all over the United States. Interest and improvements in the theory came rapidly. Here is an easy example of this procedure which I often selected: One may choose a computation of a volume of a region defined by a number of equations or inequalities in spaces of a high number of dimensions. Instead of the classical method of approximating everything by a network of points or "cells," which would involve billions of individual elements, one may merely select a few thousand points at random and obtain by sampling an idea of the value one seeks.

The first questions concerned the production of the random or pseudo-random numbers. Tricks were quickly devised to produce them internally in the machine itself without relying on any outside physical mechanism. (Clicks from a radioactive source or from cosmic rays would have been very good but too slow.) Beyond the literal or "true" imitation of a physical process on electronic computers, a whole technique began to develop on how to study mathematical equations which on their face seem to have nothing to do with probability processes, diffusion of particles, or chain processes. The question was how to change such operator equations or differential equations into a form that would allow the possibility of a probabilistic interpretation. This is one of the main theses behind the Monte Carlo method, and its possibilities are not yet exhausted. I felt that in a way one could invert a statement by Laplace. He asserts that the theory of probability is nothing but calculus applied to common sense. Monte Carlo is common sense applied to mathematical formulations of physical laws and processes.

Much more generally, electronic computers were to change the face of technology. We discussed the many possibilities endlessly. But not even von Neumann could foresee their full economic or technological impact. These aspects of their development were still in their infancy so far as industrial applications were concerned when he died in 1957. Little did we know in 1946 that computing would become a fifty-billion-dollar industry annually by 1970.

Almost immediately after the war Johnny and I also began to discuss the possibilities of using computers heuristically to try to obtain insights into questions of pure mathematics. By producing examples and by observing the properties of special mathematical objects one could hope to obtain clues as to the behavior of general statements which have been tested on examples. I remember proposing in 1946 a calculation of a very great number of primitive roots of integers so that by observing the distributions one obtained enough statistical material on their appearance and on the combinatorial behavior to perhaps get some ideas of how to state and prove some possible general regularities. I do not think that this particular program has been advanced much until now. (In mathematical exploratory work on computers my collaborators were especially Myron Stein and Robert Schrandt.) In the following years in a number of published papers, I have suggested—and in some cases solved—a variety of problems in pure mathematics by such experimenting or even merely "observing." The Gedanken Experimente, or Thought Experiments, of Einstein are possible and often useful in the purest part of mathematics. One of the papers outlining a field of exploration in "non-linear problems" was written in collaboration with Paul Stein. By now, a whole literature exists in this field.

Quite early, in fact only some months after the electronic computer called MANIAC became available in Los Alamos, I tried with a number of associates (Paul Stein, Mark Wells, James Kister, and William Walden) to code the machine to play chess. It was not so terribly difficult to code it to play correctly according to the rules. The real problem is that, even today, nobody knows how to put in its memory experiences of previous games and a general recognition of the quality of patterns and positions. Nevertheless it can be made to play so it can beat a rank amateur. We realized that the differences between playing poorly and playing well are much greater than teaching it to make legal moves and respond to obvious threats, and so on. This game was played on a six-by-six board without bishops (to shorten the time between moves). We wrote an article that appeared in the *U.S. Chess Review* and was soon reprinted by a Russian chess magazine. Stein, originally a physicist, "converted" to mathematics and became one of my closest collaborators.

Curiously the patterns of chess remind me of oriental rugs and also of something that laymen won't understand—very complicated non-measurable sets. I think I am a fair chess player. When I first came to this country I played with other mathematicians for relaxation. In Los Alamos during the years after the war, friends and younger colleagues organized a chess club, and I played many games. The Los Alamos chess team on which I played board number one several times beat Santa Fe and even Albuquerque with their populations respectively three and fifteen times that of Los Alamos.

It was in 1949, after Teller's return, that George Gamow, whom I had met briefly in Princeton before the war, came to Los Alamos for a lengthy visit. He was on a year's leave from George Washington University in Washington. In physical appearance he was quite an impressive man, six feet three inches tall, slim in 1937 (by 1949 heavy set), blond, blue-eyed, youthful looking, full of good humor. He had a very characteristic shuffling way of walking with mincing steps. He was very different from the popular picture of the specialized, scholarly scientist—not at all the standard type of academic personage. There was nothing dry about him. A truly "three-dimensional" person, he was exuberant, full of life, interested in copious quantities of good food, fond of anecdotes, and inordinately given to practical jokes.

Almost at once we became friends and engaged in interminable discussions. In some ways our temperaments matched. He found something congenial in my way of thinking (or not thinking) about problems of physics along standard lines. He liked to approach different problems from many different directions in an unassuming, direct, and original way. He talked about himself a great deal. Generally he was one of the most egocentric persons I have known, yet paradoxically (because this combination is so rare) he was at the same time completely devoid of malice towards others.

It was he (and Edward U. Condon, independently and almost simultaneously), who started theoretical nuclear physics in a 1928 paper on the quantum theoretical explanation of radioactivity. In scientific research, he concentrated on a few given problems over a period of years, returning to the same questions time and again.

Banach once told me, "Good mathematicians see analogies between theorems or theories, the very best ones see analogies between analogies." Gamow possessed this ability to see analogies between models for physical theories to an almost uncanny degree. In our ever-more-complicated and perhaps oversophisticated uses of mathematics, it was wonderful to see how far he could go using intuitive pictures and analogies from historical or even artistic comparisons. Another quality of his work was the nature of the topics with which he dealt. He never allowed his facility to carry him away from the essence of his subject in pursuit of unimportant details and elaborations. It was along the great lines of the foundations of physics, in cosmology, and in the recent discoveries in molecular biology that his ideas played an important role. His pioneering work in explaining the radioactive decay of atoms was followed by his theory of the explosive beginning of the universe, the "big bang" theory (he disliked the term by the way), and the subsequent formation of galaxies. The recent discovery of the radiation pervading the universe, corresponding to a temperature of some three degrees absolute, seems to confirm his prediction in 1948 concerning residual radiation from the big bang about ten billion years ago. This discovery came after his death in 1968.

Gamow, who was a complete layman in the field of biology (some of his detractors would say almost a charlatan), proposed, with his fantastically unerring instinct, some ideas about how the code really worked. I think he was the first to suggest that the sequence of the four substances of the DNA denoted by the letters A, C, T, G, expressed words, and how from these four letters one could build twenty or twenty-three amino acids which, in turn, considered as words, combined into phrases defining the structures of proteins. Gamow had this idea before anyone else. He even almost had the correct way (later found by Crick) of expressing the formation by triplets. At first he thought four were necessary. He was almost correct from the start.

One may see in his work, among other outstanding traits, perhaps the last example of amateurism in science on a grand scale.

An overwhelming curiosity about the scheme of things in nature, in the very large and in the very small, directed his work in nuclear physics and in cosmology.

The meaning, the origin and perhaps the variability in time of the fundamental physical constants like c (the velocity of light), h (the Planck constant), G (the gravitation constant) occupied his imagination and his efforts during the last years of his life.

The great unanswered questions concern the relations between masses of elementary particles and also the very large numbers which are the ratios between the nuclear, electrical, and gravitational forces. Gamow thought that these numbers could not have arisen as a result of an initial accident, and that they might be obtainable from topological or number-theoretical considerations. He believed in the final simplicity of a theory which one day would explain these numbers.

The French fictional detective Arsène Lupin, arch rival of Sherlock Holmes, said: *"Il faut commencer à raisonner par le bon bout."* (You have to start thinking from the right end.) Gamow had a particularly great gift for this. He used models or similes; speaking mathematically he was guided by isomorphisms or homomorphisms. Abstruse ideas of quantum theory and the more palpable ideas between structures of classical physics were transformed or transfigured, not just by repetition, but by going to variables of higher time, to use technical mathematical terms again.

In 1954 Gamow and I happened to be in Cambridge, Massachusetts, at the same time. I was telling him about some of my speculations on the problems of evolution and the possibilities of calculating the rate of evolution of life. One day he came to see me and said: "Let's go to Massachusetts General Hospital—there is an interesting biology seminar." And we drove in his Mercedes. On the way I asked him who was talking. He said, "You are!" Apparently he had told the professors running the seminar that we would both talk about these speculations. And indeed we both did. On the way home I remarked, "Imagine, George, you and me trying to talk about biology! All these people, all these doctors in white smocks—they were ready to put us in straitjackets."

In conversations during the last few months of his life he often returned to the consideration of schemata that might possibly throw light on the mystery of the elementary particles and the constants of physics. In a dream he had, which he related to his wife, Barbara, shortly before his death, he described the tantalizing experience of being in the company of such great spirits as Newton and Einstein and of discovering, as they had discovered, the extreme simplicity of the ultimate scientific truths.

At the same time that he delighted in cutting to the heart of things, he kept track of all his mundane activities in a very detailed and systematic way. From the first time I met him to the end, when we were both professors on the same campus in Boulder, I remember his collecting and putting in order all manner of snapshots and pictures of his various activities, markers as it were of scientific progress, vacation trips, discussions with friends. He also loved to compose photo-montages combining his own drawings with photo-graphic cut-outs. These were intended as illustrations or car-icatures of scientific discoveries.

All his writings are characterized by a natural flow of ideas, a simple uninvolved presentation, and an easy, never redundant, amusing but never frivolous style. He wrote eas-ily, quickly, hardly ever rewriting, filling innumerable pages, each with only a few lines handwritten in enormous characters.

His now classic books on the history of physics and on the new ideas in the physical sciences show him to have been without malice or harsh judgment towards fellow physicists. He was sparing with real praise, reserving it only for the great achievements, but he never criticized or even pointed out mediocrity.

His popular books on science received great acclaim. Among the outstanding qualities of these works are simplic-ity of approach and the avoidance of unnecessary technical details that also distinguished his work in research.

His honesty made him write exactly the way he thought, embodying the precept of Descartes: "ordering one's thoughts to analyze the complex by dissecting it into its simpler parts."

One characteristic of Gamow, which was not perhaps di-rectly visible but was easily deducible from his conversation and his creative activity, was his excellent memory. After dinner or at parties he loved to recite for the benefit of friends of Slavic origin long excerpts of Russian poetry; he could quote Pushkin or Lermontov by the hour. He also loved to use Russian proverbs.

Gamow had a ready wit and made many bons mots. He told me that the day he drove to Los Alamos for the first time, he noticed that "as one crosses the Rio Grande, and before one arrives at the Valle Grande [the Valle Grande is an extinct volcano of enormous dimensions in the mountains behind Los Alamos], one comes to the city of the Bomba Grande."

In 1949 my approaching fortieth birthday appeared as a threatening landmark in my life. I always considered it ominous to be slipping into middle age. One's feelings about age change with time, of course, but mathematicians have a reputation for peaking early, and many, myself included, have an admiration for youth. This Hellenic accent on youth is also something of an American obsession. From my earliest reading I admired Abel, the Norwegian genius who died at twenty-seven, and Galois, the creator of new ideas in algebra and group theory who died in a duel in Paris at the age of twenty-one. The greatest achievements of middle-aged men did not touch me as much. The tragedy, of course, is that as one gets older one tends to try to use old tricks in new situations; a sort of self-poisoning stops creativity. My friend Rota said he did not believe it was creativity that stopped but interest. That remark sparked a feeling of déjà vu. I agree in part. Maybe it is like boxing: it is not that the reactions slow down or that one is more easily fatigued; when boxers start having to think about what they are doing, they lose, because the reaction should be instinctive, quick automatic subroutines, so to speak.

Johnny used to say that after the age of twenty-six a mathematician begins to go downhill. When I met him he was just past that age. As time went on he extended the limit, but kept it always a little below his age. (For example, when near forty, he raised it to thirty-five.) This was characteristic of his rather self-effacing manner. He did not want to give the appearance of considering himself "in." He knew that self-praise sounds ridiculous to others, and he would lean over backwards to appear modest. I, on the contrary, always took pleasure in boasting, especially about some of my own trivial accomplishments like athletics or winning at games. Children boast quite naturally. In the literature of antiquity, notably in Homer, heroes brag openly about their athletic prowess. Scientists sometimes boast by implication when they criticize or minimize the achievements of others.

Approaching forty and thinking about what I had accomplished up to that point, I was still very hopeful that much work lay ahead of me. Perhaps because much of what I had worked on or thought about had not yet been put in writing, I felt I still had things in reserve. Given this optimistic nature, I feel this way even now when I am past sixty.

CHAPTER 11

The "Super"
1949–1952

I WAS returning to Los Alamos from one of my frequent trips east when our monitoring systems detected the Russian A-bomb explosion. The news had not yet been made public. Immediately upon my arrival several friends—Metropolis, Calkin, and others—who met me at the little airport, greeted me with these items of news: (a) in a poker game the night before, Jack had won eighty dollars (an enormous sum for our kind of stakes), and (b) the Russians had detonated an atomic bomb. I considered this for a moment and said, of course, I believed (a). It was (b) that was true.

Johnny was in Los Alamos and he and Teller had been spending time together discussing this ominous development. I joined them in Johnny's room at the Lodge. The general question was "What now?" At once I said that work should be pushed on the "super." Teller nodded. Needless to say, that was on his mind also. They said they had been discussing how to go about it. The next day Teller left for

Washington, perhaps to see Admiral Strauss, who was a member of the Atomic Energy Commission, to do what politicking he could there.

Strauss was one of the first AEC Commissioners. He was Jewish. Talking to him I noticed that he had the rather common—and to me pleasant—Jewish tendency to admire successful scientists. He possessed a sort of wistful appreciation of science, perhaps because he was not a scientist himself. In his early days on the Commission he had pushed for the development of a monitoring system to detect the presence of nuclear work anywhere in the world. This could be done by examining air samples from the atmosphere for the presence of certain gases which came from uranium fission. The idea came from Tony Turkevitch, a physical chemist from Chicago. I remember his mentioning such a plan in my presence in Los Alamos during the war.

To what extent Strauss's counsel, influenced by Teller and von Neumann, contributed to President Truman's decision to order full-speed work on the H-bomb, I do not know.

It has to be repeated here again that the work on the "super" had been going on efficiently and systematically. Norris Bradbury was directing the allocation of the theoretical effort. Some six months before this news from Russia, I mentioned to him that I had the impression that some people in Washington did not want this work to continue and Norris had said: "I'll be damned if I'll let anybody in Washington or any politicians tell me what work not to do." I remember his smiling expression when he said it. This sentiment was not what is now called hawkish or motivated by political or military considerations; it referred purely to scientific and technological inquiry.

Theoretical work on the "super" had, as I have shown, continued all along after the "super" conference, but I do not think Teller wanted this publicized very much, for I think he either believed he was or wanted to be known as not only the main but the sole promoter, defender, and organizer of the work. Perhaps he felt that Bradbury, as head of the laboratory, would receive most of the credit for the future thermonuclear bomb, just as Oppenheimer had for the A-bomb, to the exclusion of the other scientists who had done the technical work. Indeed Teller had been and was the original proponent of intensive work on thermonuclear explosions in the United States.

This, of course, is my own interpretation of the reasons for the developments that followed. It may be substantiated in the existing literature, including the notorious Shepley-Blair account published in *Life* magazine, which contributed so much to the establishment of Teller as the "Father of the H-Bomb." Their subsequent book was later discredited because of the misinformation it contained.

Shortly after President Truman's announcement directing the AEC to proceed with work on the H-Bomb, E. O. Lawrence and Luis Alvarez visited Los Alamos from Berkeley and started discussions with Bradbury and then with Gamow, Teller, and myself about the feasibility of constructing a "super." This visit played a part in the politics of this enterprise.

One of Teller's first moves was to enlist a young physicist, Frédéric de Hoffmann, as his assistant. A native of Vienna, Freddy had come to the United States as a young boy before the war. He was young, clever, intelligent, and quick, but not what you might call a really original scientist. He became a sort of factotum, a jack of all trades for communications, contacts with administrators, and other duties. He carried Edward's messages back and forth to Washington and also did some technical work. He was an ideal associate for Edward, who later could afford to be generous with credit for Freddy's contributions, which would not detract from his own appearance as the almost exclusive originator, propagator, and executor of the project.

A first committee was formed to organize all work on the "super" and investigate all possible schemes for constructing it. The committee's work was directed by Teller, as chairman, Gamow, and myself.

Several different proposals of ideas existed on how to initiate the thermonuclear reaction, using fission bombs as starter. One of Gamow's was called "the cat's tail." Another was Edward's original proposal. Gamow drew a humorous cartoon with symbolic representations of these various schemes. In it he squeezes a cat by the tail, I spit in a spittoon, and Teller wears an Indian fertility necklace, which according to Gamow is the symbol for the womb, a word he pronounced "vombb." This cartoon has appeared among the illustrations in his autobiography, *My World Line*, published by The Viking Press in 1970.

Both Gamow and I showed a lot of independence of thought in our meetings, and Teller did not like this very much. Not too surprisingly, the original "super" directing committee soon ceased to exist. At a moment when both Gamow and I were out of town, Teller prevailed upon Bradbury to disband the committee and to replace it by another organizational entity. Gamow was quite put out by this. I did not care, but I wrote him, prophetically it seems, that great troubles would follow because of Edward's obstinacy, his single-mindedness, and his overwhelming ambition. This letter, like all communications about work in Los Alamos, was "classified." I expect it is still filed somewhere and perhaps some day may be included in some collection of documents from that period. Another such "indiscreet" letter is one I wrote to von Neumann in which I made fun of Edward's attitude. This letter happens to be quoted in the second volume of the official history of the AEC, *The Atomic Shield.* In it I mention that an idea occurred to me which I had communicated to Edward. I added in jest that since Edward liked it very much perhaps that meant it would not work either.

At some time I outlined a possible detailed calculation which became the base of the work carried out by von Neumann on the newly built electronic computing machines with the help of Klari, as a programmer, and Cerda and Foster Evans, a husband-and-wife team of physicists who had joined the project after the war.

The wartime, or rather, immediate postwar Metropolis-Fraenkel calculation was very schematic compared to what I had in mind. More ambitious calculations of this sort had become possible since computers had improved both in speed and in the size of their memories. The steps outlined involved a fantastic number of arithmetical operations. Johnny said to me one day, "This computation will require more multiplications than have ever been done before by all of humanity." But when I estimated roughly the number of multiplications performed by all the world's school children in the last fifty years, I found that this number was larger by about a factor of ten!

Ours was the biggest problem ever, vastly larger than any astronomical calculation done to that date on hand computers, and it needed the most advanced electronic equipment available. By then von Neumann's Princeton MANIAC was functioning and a duplicate version was being built in Los Alamos under the direction of Metropolis.

Still Teller kept on hinting that not enough work was being done on his original scheme for the ignition of the "super." He kept insisting on certain special approaches of his own. I must admit that I became irritated by his insistence; in collaboration with my friend Everett one day I decided to try a schematic pilot calculation which could give an order of magnitude, at least, a "ballpark" estimate of the promise of his scheme.

Before we started this calculation of the progress of a thermonuclear reaction (burning in a mass of deuterium or deuterium-tritium mixture), Everett and I had done a lot of work on probability questions connected with the active assemblies of uranium and with neutron multiplications. We worked out a theory of multiplicative processes, as we called it. (Now the preferred name is "branching processes.") This work followed the ideas developed in the report on branching processes written with David Hawkins during the war. But it elaborated, deepened, and extended them very greatly. The report of Hawkins and myself conisted of a few pages. The results of several months' work with Everett were contained in three large reports of a hundred pages or more. The latter became the basis for much subsequent work, some of it done later independently by Russian and Czech scientists.

Within the formal organization of the laboratory, I was the leader of Group T-8, and Everett was its only member. Every day, in his office adjoining mine, we had done quite a lot of other mathematical work not necessarily connected with the current programmatic questions of Los Alamos and we would discuss the "universe" (Johnny's expression), mathematical or otherwise.

Now we started to work each day for four to six hours with slide rule, pencil, and paper, making frequent quantitative guesses, and managed to get approximate results much more quickly than the gigantic problem, which was progressing slowly. Much of our work was done by guessing values of geometrical factors, imagining intersections of solids, estimating volumes, and estimating chances of points escaping. We did this repeatedly for hours, liberally sprinkling the guesses with constant slide-rule calculations. It was long and arduous work, the results of which gave a rather discouraging picture of the feasibility of the original "super" scheme. Our calculation showed its enormous practical difficulties and threw grave doubts on the prospects of Edward's original approach to the initial ignition conditions of the "super."

We proceeded something like this: each morning I would attempt to supply several guesses as to the value of certain coefficients referring to purely geometrical properties of the moving assembly involving the fate of the neutrons and other particles going through it and causing, in turn, more reactions. These estimates were interspersed with stepwise calculations of the behavior of the actual motions. The reader should realize that the real times for the individual computational steps were short, each less than a "shake," and the spatial subdivisions of the material assembly very small in linear dimensions. Each step took a fraction of a "shake." A "shake" was the name given in Los Alamos during the war to the time interval of 10^{-8} seconds. Another unit was that of a cross section called a "barn"; it was 10^{-24} of a square centimeter, a terribly small area. The number of individual computational steps was therefore very large. We filled page upon page of calculations, much of it done by Everett. In the process he almost wore out his own slide rule, and when our results were achieved, after several months of work, Everett joked that "the grateful government could at least offer to buy him a new slide rule." I do not know how many man-hours were spent on this problem.

To write the reports we enlisted the help of professional computers, Josephine Elliott among them. Even Françoise was pressed into service, grinding out untold numbers of arithmetical operations on desk calculators.

Lengthy as this process was, the work was finished several months before the Princeton electronic computer's results started coming in. This so-to-speak homespun part of the H-bomb development was described in many official and popular accounts. It caught the public eye probably because of the certain appeal of its "man versus machine" element.

As our calculation progressed it naturally attracted quite a lot of attention among the physicists Teller was trying to interest in the "super" project, and also among those Bradbury had already enlisted for this work. Distinguished visitors would appear periodically to see how the calculations were going. John Wheeler's first visits to Los Alamos date from about this period.

One day Fermi and Rabi came to our office, and we showed them the results which pointed to the mediocre progress of the reaction. These results could only be indicative and were by no means certain because of the crude approximations and guesses which we used in the place of voluminous numerical operations.

When it appeared that the technical difficulties of Teller's original ideas could justify some of the scientific and political objections of certain physicists, and even perhaps the reluctance of the General Advisory Committee, Hans Bethe evidenced a renewed interest in the whole project and came to visit Los Alamos more often. With his wonderful virtuosity in mathematical physics and with his ability to solve analytical problems of nuclear physics he helped significantly. After all, it was Bethe who first suggested (Weizsäcker in Germany had reached the same conclusion independently) that nuclear reactions in the interior of the sun could be responsible for the sun's energy generation and thus explain the radiation emitted by the sun and by other stars. Their original "carbon reaction mechanism" has been found not to be quite so exclusively responsible for the energy generation in all stars as was originally thought.

Teller was not easily reconciled to our results. I learned that the bad news drove him once to tears of frustration, and he suffered great disappointment. I never saw him personally in that condition, but he certainly appeared glum in those days, and so were other enthusiasts of the H-bomb project. Subdued and depressed, he would visit our offices periodically and would attempt to prove us wrong by trying to find mistakes. Once he said, "There is a mistake here by a factor 10^4." This especially annoyed Everett who did not have much self-confidence as a physicist but who, as a mathematician, amazingly enough never made mistakes. He used to say, "I never make mistakes" and this was true in that he never used a wrong sign or made simple numerical mistakes, as mathematicians often do. But each time he tried, Edward had to admit that it was he who was at fault in his arithmetic.

As the results of the von Neumann–Evans calculation on the big electronic Princeton machine slowly started to come in, they confirmed broadly what we had shown. There, in the course of the calculation, in spite of an initial, hopeful-looking "flare up," the whole assembly started slowly to cool down. Every few days Johnny would call in some results. "Icicles are forming," he would say dejectedly.

These computations were the best one could do theoretically in those days. Because the existing experimental values for the constants which had to be applied in the calculations for the cross sections were uncertain, the project was still alive, but it was necessary to search for alternative approaches to ignition.

All along Johnny was emotionally involved in favor of the construction of an H-bomb. He hoped that in one way or another a good scheme would be found, and he never lost heart even when the mathematical results for the original approach were negative.

During this crucial period of uncertainty, I visited him in Princeton. Fermi happened to be there, too, on a brief visit, and we discussed the prospects all afternoon, during dinner in Johnny's house, and all evening. The next day we talked with Oppenheimer. He knew about the results obtained by Everett and me. He seemed rather glad to learn of the difficulties, whereas von Neumann was still searching for ways to rescue the whole thing. Johnny outlined some hydrodynamical calculations. Fermi concurred. They came to estimate a certain velocity of expansion which seemed to me much too slow. With the experience of all the work I had done in the past months, I noticed that they had erred in assuming the density of liquid deuterium to be 1, whereas it is only a small fraction of 1. This error of per unit mass instead of per unit volume made the velocity appear indeed smaller and Johnny realized it and exclaimed: "Oh my! It is indeed much faster than a train." Oppenheimer winked at me. He liked having the difficulties confirmed and enjoyed catching von Neumann and Fermi in a small and trivial arithmetic error.

My calculations with Everett concerned the first phase of

the explosion, the problem of the initial ignition. An important part of the story has been overlooked in the official accounts and concerns some quite fundamental work that Fermi and I did following the first calculation of the progress of the reaction, its propagation and explosion. In numerous joint discussions we outlined the possibilities of propagation, assuming that some way or other (perhaps by the expenditure of large amounts of tritium) the initial ignition could be achieved. There again we had to use guesses in place of the enormously difficult detailed calculations that would have required computers even faster than those in existence. We did this again in time-step stages with intuitive estimates and marvelous simplifications introduced by Fermi.

The numerical work was done on desk computers with the assistance of a number of programmers from the laboratory's computing group, managed by a good-humored New Yorker, Max Goldstein. Much to Max's annoyance, Fermi wanted to encourage the girls to use slide rules; the machine precision was not really warranted because of our simplifications. But Max insisted on the usual routines with desk calculators. Reading from slide rules and using logarithms as Fermi did was much less accurate, but with his marvelous sense he had the ability to judge the right amount of accuracy which would be meaningful. The girls, who were merely making calculations without knowing the physics or general mathematics behind them, could not do that, of course, so in a way Max was quite right in insisting on the standard routines.

I particularly remember one of the programmers who was really beautiful and well endowed. She would come to my office with the results of the daily computation. Large sheets of paper were filled with numbers. She would unfold them in front of her low-cut Spanish blouse and ask, "How do they look?" and I would exclaim, "They look marvelous!" to the entertainment of Fermi and others in the office at the time.

A joint report was written by Fermi and myself. Enrico exercised very great caution in its conclusions. In fact, one conclusion stating the unpromising nature of the reaction as it was planned contained this sentence: "If the cross sections for the nuclear reactions could somehow be two or three times larger than what was measured and assumed, the reaction could behave more successfully."

I believe this work with Fermi to have been even more important than the calculations made with Everett. It turned out to be basic to the technology of thermonuclear explosions. Fermi was satisfied with both its execution and with the fact that it put a limit to the size of such explosions. As he said: "One cannot make trees grow skyward indefinitely."

In the meantime Teller continued to be very active both politically and organizationally at the moment when things looked at their worst for his original wartime "super" design, even with the modifications and improvements he and his collaborators had outlined in the intervening period.

Perhaps the change came with a proposal I contributed. I thought of a way to modify the whole approach by injecting a repetition of certain arrangements. Unfortunately, the idea or set of ideas involved is still classified and cannot be described here.

Psychologically it was perhaps precipitated by a memorandum from Darol Froman, an associate director of the laboratory, who asked various people what should be done with the whole "super" program. While expressing doubts about the validity of Teller's insistence on his own particular scheme, I wrote to Froman that one should continue at all costs the theoretical work, that a way had to be found to extract great amounts of energy from thermonuclear reactions.

Shortly after responding I thought of an iterative scheme. After I put my thoughts in order and made a semi-concrete sketch, I went to Carson Mark to discuss it. Mark, who was by then head of the theoretical division, was already in charge of the very extensive theoretical work supporting Teller's and Wheeler's special groups. The same afternoon I went to see Norris Bradbury and mentioned this scheme. He quickly grasped its possibilities and at once showed great interest in pursuing it. The next morning, I spoke to Teller. I don't think he had any real animosity toward me for the negative results of the work with Everett so damaging to his plans, but our relationship seemed definitely strained. At once Edward took up my suggestions, hesitantly at first, but enthusiastically after a few hours. He had seen not only the novel elements, but had found a parallel version, an alternative to what I had said, perhaps more convenient and generalized. From then on pessimism gave way to hope. In the following days I saw Edward several times. We discussed the problem for about half an hour each time. I wrote a first sketch of the proposal. Teller made some changes and additions, and we wrote a joint report quickly. It contained the first engineering sketches of the new possibilities of starting thermonuclear explosions. We wrote about two parallel schemes based on these principles. The report became the fundamental basis for the design of the first successful thermonuclear reactions and the test in the Pacific called "Mike." A flurry of activity ensued. Teller lost no time in presenting these ideas, perhaps with most of the emphasis on the second half of our paper, at a General Advisory Committee meeting in Princeton which was to become quite famous because it marked the turning point in the development of the H-bomb. A more detailed follow-up report was written by Teller and de Hoffmann. New physicists were brought to Los Alamos, and work toward experimental verification started in earnest.

John Wheeler came to New Mexico to help Teller. He brought with him several of his brightest students. Among them was Ken Ford, with whom I was to do unrelated work later on. There was John Toll, now President of the State University of New York at Stony Brook, then a promising young physicist; Marshall Rosenbluth, who had been in Los

Alamos during the war as a soldier with the SED; Ted Taylor, who contributed so many new ideas to atomic fission bombs; a brilliant mathematical physicist, Conrad Longmire; and other talented young people. Intense, fast work went on, and the plans for "Mike" were ready just a few months after our fateful conversation.

During their Los Alamos year, the Wheelers lived in a house next to ours and we saw them frequently. Wheeler was a very interesting type of physicist. To my mind, he had all the right desires for novelty in theoretical ideas without sticking too rigidly to preconceived notions and existing schemes. Sometimes he thought of outlandish-sounding schemes in physics or in cosmology, so much so that some of his ideas would strike me as lacking in an element of common sense or connection with possible experiments. Or perhaps they were, on the contrary, as Pauli once said about some ideas proposed by Heisenberg, "not crazy enough." Wheeler's great merit is his work on general relativity, pursued to extreme situations like black holes and beyond; he also has great didactical talents. Of his students, the best by far, I think, is Feynman. Long before, they wrote a very nice joint paper on a generalization of Mach's principle.

Despite the great flurry of excellent experiments and the thermonuclear explosion itself, Teller continued to be dissatisfied and engaged in multiple activities in an effort to put still more of the work under his control. He expressed great unhappiness with the way Los Alamos handled the developments, though Bradbury and other senior members of the laboratory could see no other rational way of doing things. The rift grew so large that Teller put on all the political pressure he could muster to start a rival laboratory. Thanks largely to his influence with Lewis Strauss and the Commission in Washington, he obtained funds and authorization to start and staff another laboratory in Livermore, California, at about the time of the very successful "Mike" test which more than confirmed the possibilities. So Los Alamos went on to build the first H-bomb without him, while some of the first designs emanating from Livermore were quite unsuccessful. Johnny was aware of the feeling between the two laboratories. After the first unsuccessful Livermore try at a thermonuclear explosion in the Pacific proving grounds he laughed and said to me: "There will be dancing in the streets of Los Alamos tonight."

Contrary to those people who were violently against the bomb on political, moral or sociological grounds, I never had any questions about doing purely theoretical work. I did not feel it was immoral to try to calculate physical phenomena. Whether it was worthwhile strategically was an entirely different aspect of the problem—in fact the crux of a historical, political or sociological question of the gravest kind—and had little to do with the physical or technological problem itself. Even the simplest calculation in the purest mathematics can have terrible consequences. Without the invention of the infinitesimal calculus most of our technology would have been impossible. Should we say therefore that calculus is bad?

I felt that one should not *initiate* projects leading to possibly horrible ends. But once such possibilities exist, is it not better to examine whether or not they are real? An even greater conceit is to assume that if you yourself won't work on it, it can't be done at all. I sincerely felt it was safer to keep these matters in the hands of scientists and people who are accustomed to objective judgments rather than in those of demagogues or jingoists, or even well-meaning but technically uninformed politicians. And when I reflected on the end results, they did not seem so qualitatively different from those possible with existing fission bombs. After the war it was clear that A-bombs of enormous size could be made. The thermonuclear schemes were neither very original nor exceptional. Sooner or later the Russians or others would investigate and build them. The political implications were unclear despite the hullabaloo and exaggerations on both sides. That single bombs were able to destroy the largest cities could render all-out wars less probable than they were with the already existing A-bombs and their horrible destructive power.

After completing this theoretical work I considered my job done and decided to change surroundings for a while. I accepted an invitation to spend a semester at Harvard as a visiting professor. It was the summer of 1951. The Fermis lived in the other half of our duplex. We saw them often. In September, as I was preparing to leave, I was packing and working on correspondence, books and papers and forgot to attend an important evening meeting in Bradbury's office on the planning of future work and experiments. The next morning I learned that there had been a series of heated exchanges between Teller and Bradbury, and that some acrimonious remarks by other scientists present had been directed at Teller's rather wild accusations. As I commented on this to Fermi, he replied with his usual serene imperturbability: "Why should you care, you are going away the day after tomorrow." Some of my friends were greatly impressed by this display of olympian detachment. Rabi in particular admired Enrico's logically calm attitudes.

The Oppenheimer Affair, which grew out of the violent hydrogen-bomb debate–even though the animosity between Strauss and Oppenheimer had personal and perhaps petty origins—did greatly affect the psychological and emotional role of scientists.

Once I asked Johnny whether he thought that Einstein would have actively defended Oppenheimer during the latter's troubles. Johnny replied that he believed not; he thought Einstein had had genuinely mixed feelings about some of Oppenheimer's actions and about the Affair.

It is hard to guess another's motives. They may be the result of long-held convictions, political orientation, or even pet scientific or philosophical ideas. I believe, for example, that perhaps some of the reasons for Oppenheimer's opposition to the development of the H-bomb were not exclusively on moral, philosophical, or humanitarian grounds. I might say cynically that he struck me as someone who, having been instrumental in starting a revolution (and the advent of nuclear energy does merit this appellation), does not contemplate with pleasure still bigger revolutions to come.

Anatole France tells somewhere that one day in a park in Paris he saw an old man sitting on a bench reading a newspaper. Suddenly a group of young students appeared, marching in parade formation and shouting revolutionary slogans. The old man became very agitated, shaking his cane and shouting: "Order! Police! Police! Stop!" France recognized the old man; in the past he had been a famous revolutionary.

Oppenheimer had many unusually strong, interesting qualities; but in some way he was a very sad man. The theoretical discussion which he proposed of the so-called neutron stars is one of his great contributions to theoretical physics, but its verification with the discoveries of pulsar stars, which are fast-rotating neutron stars, came years after his death.

It seems to me this was the tragedy of Oppenheimer. He was more intelligent, receptive, and brilliantly critical than deeply original. Also he was caught in his own web, a web not of politics but of phrasing. Perhaps he exaggerated his role when he saw himself as "Prince of Darkness, the destroyer of Universes." Johnny used to say, "Some people profess guilt to claim credit for the sin."

Many accounts of these events have been written. Some are exaggerated or distorted; others, like the official history of the AEC, are rather objective. But none can be complete yet, and of course the events as seen by the participants appear in different lights. This is my own account of the history of the H-bomb as I lived it and to the extent that I was directly involved in it.

The Death of Two Pioneers

1952–1957

AFTER the somewhat frantic work on the "super" first with Everett then with Fermi, and my return from the semester's leave of absence at Harvard renewing contacts with old mathematical friends, my preoccupations turned to other and more purely scientific problems.

Computers were brand-new; in fact the Los Alamos MANIAC was barely finished. The Princeton von Neumann machine had met with technical and engineering difficulties that had prolonged its perfection. The Los Alamos model had been luckier, for it was in the capable hands of James Richardson, an engineer in the Metropolis group.

As soon as the machines were finished, Fermi, with his great common sense and intuition, recognized immediately their importance for the study of problems in theoretical physics, astrophysics, and classical physics. We discussed this at length and decided to attempt to formulate a problem simple to state, but such that a solution would require a lengthy computation which could not be done with pencil and paper or with the existing mechanical computers. After deliberating about possible problems, we found a typical one requiring long-range prediction and long-time behavior of a dynamical system. It was the consideration of an elastic string with two fixed ends, subject not only to the usual elastic force of strain proportional to strain, but having, in addition, a physically correct small non-linear term. The question was to find out how this non-linearity after very many periods of vibrations would gradually alter the well-known periodic behavior of back and forth oscillation in one mode; how other modes of the string would become more important; and how, we thought, the entire motion would ultimately thermalize, imitating perhaps the behavior of fluids which are initially laminar and become more and more turbulent and convert their macroscopic motion into heat.

John Pasta, a recently arrived physicist, assisted us in the task of flow diagramming, programming, and running the problem on the MANIAC. Fermi had decided to try to learn how to code the machine by himself. In those days it was more difficult than now, when there are set rules, ready-made programs, and the procedure itself is automated. One had to learn many little tricks in those early days. Fermi learned them very quickly and taught me some, even though I already knew enough to be able to estimate what kind of problems could be done, their duration in number of steps, and the principles of how they should be executed.

Our problem turned out to have been felicitously chosen. The results were entirely different qualitatively from what even Fermi, with his great knowledge of wave motions, had expected. The original objective had been to see at what rate the energy of the string, initially put into a single sine wave (the note was struck as one tone), would gradually develop higher tones with the harmonics, and how the shape would finally become "a mess" both in the form of the string and in the way the energy was distributed among higher and higher modes. Nothing of the sort happened. To our surprise the string started playing a game of musical chairs, only between several low notes, and perhaps even more amazingly, after what would have been several hundred ordinary up and down vibrations, it came back almost exactly to its original sinusoidal shape.

I know that Fermi considered this to be, as he said, "a minor discovery." And when he was invited a year later to give the Gibbs Lecture (a great honorary event at the annual American Mathematical Society meeting), he intended to talk about this. He became ill before the meeting, and his lecture never took place. But the account of this work, with Fermi, Pasta and myself as authors, was published as a Los Alamos report.

I should explain that the motion of a continuous medium like a string is studied on a machine by imagining the string to be composed of a finite number of particles—in our case, sixty-four or one hundred twenty-eight. (It is better to take a power of two for the number of elements, which is more convenient to handle on the computer.) These particles are connected to each other by forces which are not only linear in terms of their distance but by additional small non-linear quadratic terms. Then the machine quickly computes in short time-steps the motion of each of these points. After having computed this, it goes to the next time-step, computes the new positions, and so on for many times. There is absolutely no way to perform this numerical work with pencil and paper; it would literally take thousands of years. An analytic closed-form solution using the mathematical techniques of classical analysis of the nineteenth and twentieth centuries is also completely unlikely.

The results were truly amazing. There were many attempts to find the reasons for this periodic and regular behavior, which was to be the starting point of what is now a large literature on non-linear vibrations. Martin Kruskal, a physicist in Princeton, and Norman Zabuski, a mathematician at Bell Labs, wrote papers about it. Later, Peter Lax contributed signally to the theory. They made interesting mathematical analyses of problems of this sort. A mathematician will know that the so-called Poincaré return-type of dynamical system containing that many particles is terrifically long—on an astronomical scale—and the fact that it came back so soon to its original form is what is so surprising.

Another Los Alamos physicist, Jim Tuck, was curious to see if after this near return to the original position, another period started again from this condition and what it would be after a second "period." With Pasta and Metropolis, he tried it again and, surprisingly, the thing came back, a percent or so less exactly. These continued and, after six or twelve such periods, it started improving again and a sort of superperiod appeared. Again this is most peculiar.

Other authors, among them several Russian mathematicians, have studied this problem and written papers about it. Last year I received a request from the Japanese Academy for permission to reprint the Fermi-Pasta-Ulam paper. I assented without hesitation and shortly thereafter a whole volume appeared containing studies of these questions by many authors.

I might say here that John Pasta was a very interesting person. A physicist by profession, he spent several years during the depression as a policeman on a beat in New York City. He joined my group in Los Alamos. On the whole very taciturn, he could occasionally make very caustic, humorous remarks. When Johnny became an AEC Commissioner, impressed by Pasta's common sense, ability, and knowledge of the Los Alamos scene, he invited him to join the AEC in Washington.

As for James Tuck, he was a British physicist who had come to Los Alamos with the British Mission during the war. He had returned to Oxford after the war, but then came back to join the laboratory again. We collaborated on a method for obtaining energy from fusion in a non-explosive way and during the war had written a joint report on this which may still be classified.

As a very young physicist Tuck was for a time assistant to Lindemann, who later became Lord Cherwell, Churchill's science advisor. He has a fund of interesting and amusing stories about this experience, and he still defends Cherwell vigorously against accusations or criticisms. He reminds me of the English eccentrics described by Jules Verne and by Karl May. Very tall, moving in an abrupt, somewhat uncoordinated way, by his awkwardness he causes many amusing incidents that are always the delight of his friends. For many years Tuck directed a Los Alamos program for the peaceful uses of fusion. The laboratory is still vigorously engaged in a large effort to find methods to extract energy "peacefully" from the fusion of deuterium.

There was another problem which Fermi wanted to study but which we somehow never came to formulate well or to work on. He said one day, "It would be interesting to do something purely kinematical. Imagine a chain consisting of very many links, rigid, but free to rotate around each other. It would be curious to see what shapes the chain would assume when it was thrown on a table, by studying purely the effects of the initial energy and the constraints, no forces."

During these years we and the von Neumanns started the practice of spending Christmas together. Claire, our daughter, was a small child, and it became a tradition that on Christmas Eve Johnny and Klari would help us assemble her toys. I remember a large cardboard dollhouse which took many hours; both Johnny and I, especially I, being inept at following the instructions which called for inserting tab A into slot B. To this day I am incapable of following written instructions, whether for filling out forms or assembling parts. Johnny, on the other hand, loved it. In Princeton he actively followed the smallest details of the construction of the Princeton MANIAC. According to Bigelow, its engi-

neer, Johnny had learned all the electronic parts and super-
vised their assembly. When the machine neared completion,
I remember how once he made fun of it at his own expense.
He told me, "I don't know how really useful this will be.
But at any rate it will be possible to get a lot of credit in
Tibet by coding 'Om Mane Padme Hum' [Oh, thou flower
of lotus] a hundred million times in an hour. It will far ex-
ceed anything prayer wheels can do."

Another Christmas we spent together was that of 1950.
To celebrate the end of the decade and the first half of the
century, Françoise, Claire, and I took a short vacation with
the von Neumanns at Guayamas, Mexico. They drove all the
way from Princeton, and we arranged to meet in Las Cruces
in the southern part of New Mexico to continue the trip
together. Las Cruces had an old 1890 brothel which had
been remodeled into a hotel after the war, and we all stayed
there. The rooms were furnished in period style and on each
door, instead of a room number, the name of a girl was inscribed: Juanita, Rosalia, Maria. In the middle of the lobby,
a swing was suspended from the ceiling. The ladies apparently climbed on it from the interior balcony. To Johnny
and me it looked like the well-known Foucault pendulum,
and we indulged in a learned and improper bilingual joke
which I will refrain from repeating.

On the drive to Guayamas we also amused ourselves with
the developing of a language which we called "neo-cas-
tillian." In our ignorance of Spanish, it consisted of English
words with Latin endings, for example el glaso, for glass. To
our great surprise and fun we found out that it worked for
some words. Terry's guide to Mexico, particularly its prose,
provided us with many hours of fun too. There was an especially eloquent page about the enchanted "paradise bosque"
of Sonora, which on driving across we found to be a miserable grove of trees on a dry sandy terrain, unoccupied by any
"numerous and diverse exotic tropical birds." This became
a proverbial expression denoting disappointment. Whenever
we heard something which did not quite measure up to ex-

pectations either in mathematics or physics, we exchanged knowingly the words "bosque encantado."

Long before Sputnik, around 1951 or 1952 I attended an early ICBM rocketry meeting in Washington. Altogether there must have been twenty or more people present. Gamow was one of the important participants. Johnny and Teller were there, too. It was a classified meeting in one of the rooms in the Pentagon. Johnny was sitting next to me at a long table. One problem under discussion was on how to guide rockets. Teller suggested a chemical path to a target. Gamow called it "smelling" the way. Other people suggested other schemes. I proposed "ballistic" projectiles, whose trajectory could be corrected if need be several times along the way. I remember Johnny asking me, "Why isn't it just as good or better to aim well from the starting point?" I reminded him of Gauss's famous work on planetary orbits calculated from several observations. He quickly thought about this for a few minutes and came to the conclusion that indeed this was a superior method.

I also noticed that at my mention of ballistic projectiles some people made embarrassed noises and I guessed that there was already some work going on on this. People would not disclose everything that was being worked on and clearances were not uniform among the participants. This brings me to a point about von Neumann that seems to have puzzled many people. It concerns his relations with the military. He seemed to admire generals and admirals and got along well with them. Even before he became an official himself, an AEC commissioner, he spent an increasing amount of time in consultation with the military establishment. Once I asked him: "How is it, Johnny, that you seem to be so impressed by even relatively minor officers who sometimes are not so very remarkable?" And I added, in order to say something a little derogatory about myself too, that what impressed me more were symbols of wealth and influence, like the sight of J. P. Morgan marching in an alumni procession at the Harvard Centennial Ceremonies in

1936. I had seen very many wonderful and eminent scientists and artists in my life before, but the sight of this man who was a billionaire and wielded enormous power really awed me. But to go back to Johnny's fascination with the military, I believe it was due more generally to his admiration for people who had power. This is not uncommon with those whose life is spent in contemplation. At any rate, it was clear that he admired people who could influence events. In addition, being softhearted, I think he had a hidden admiration for people or organizations that could be tough and ruthless. He appreciated or even envied those who at meetings could act or present their views in a way to influence not only others' thoughts, but concrete decision-making. He himself was not a very strong or active debater in committee meetings, yielding to those who insisted more forcefully. On the whole he preferred to avoid controversy.

These were the days of defense research contracts. Even mathematicians frequently were recipients. Johnny and I commented on how in some of their proposals scientists sometimes described how useful their intended research was for the national interest, whereas in reality they were motivated by bonafide scientific curiosity and an urge to write a few papers. Sometimes the utilitarian goal was mainly a pretext. This reminded us of the story of the Jew who wanted to enter a synagogue on Yom Kippur. In order to sit in a pew he had to pay for his seat, so he tried to sneak in by telling the guard he only wanted to tell Mr. Blum inside that his grandfather was very ill. But the guard refused, telling him: "Ganev, Sie wollen beten" ["You thief! You really want to pray"]. This, we liked to think, was a nice abstract illustration of the point.

Gamow, who lived in Washington, was a consultant at the Naval Research Laboratory. One of my early so-called business trips to Washington involved a consultation with him. He asked me to talk about Monte Carlo and we discussed modeling land-battle situations. He was interested in and did a lot of work on tank battles. He used Monte Carlo, for example, to simulate landscapes, which he dubbed Stan-scapes.

He lived in the suburbs with his first wife Rho and would say, "Let us meet at Chebyshev Circle." Of course, he meant Chevy Chase. (Chebyshev was a Russian mathematician, and this is how he pronounced it.) Gradually he and Rho had increasing marital difficulties and finally separated and divorced. He moved to the spartan surroundings of the Cosmos Club, where the only good thing was the profusion of newspapers and magazines available to the members. One day I received a sad letter from him saying he was living alone and that on his house was a sign saying, "For Sell [*sic*]."

In 1954 the Fermis were spending the summer in Europe, partly at the French Physics Institute in Les Houches near Chamonix, partly at Varenna in Italy, where the Enrico Fermi Institute was founded after his death. It now holds conferences on current topics in high-energy physics and in particle physics, both fields just beginning towards the end of Fermi's life.

If I recall correctly Fermi had applied for a research grant that summer and had not obtained it, which irritated him somewhat. This seemed very strange to me. Just like the incredible affair of the government's niggardly compensation for the use of his patent on the manufacture of isotopes. He told me once that he believed he and his collaborators would receive perhaps some ten million dollars from the government. With this money they wanted to establish a fund for Italians to study in the United States. But at that time they still had not received "a red cent," as he said. Eventually a settlement was made but it was so small that it barely covered the lawyers' fees, if I remember correctly.

We arranged to meet the Fermis in Paris, where they were to stay a few days, and to drive together in two cars part of the way south. They planned to rent a small Fiat but the director of Fiat in Paris made a great point of giving them a very special eight-speed car. I remember Fermi inviting me to try the car with all its speed changes along the quais and the Rue de Rivoli.

Enrico's health was by now not the best. The preceding summer in Los Alamos, his wife Laura had noticed that his appetite was poor, and this began to worry her. He also showed less energy for exercise and in the tennis games he loved to play. But there were no other physical symptoms, and Laura thought that this was possibly due to his involvement in the H-bomb controversy and the Oppenheimer affair, and to his skepticism and pessimism about the general state of world politics. She hoped the summer's rest away from home would do him good.

The Fermis always lived simply and frugally, and in Paris we noticed that they were reluctant to frequent the "good" and expensive French restaurants. Enrico did not really enjoy food that summer. Neither could we persuade them to stay in first-class hotels, which they certainly could afford better than we could in those days. On our little overnight joint trip we followed suit and took simple lodgings with them in a modest, small inn in the Vallée du Cousin some hundred fifty miles south of Paris. It is curious how one remembers physical settings. It was late evening, there were stars in the sky and we sat on a terrace next to a murmuring stream discussing the Oppenheimer Affair. Some electrical wires were strung between two houses and all the while we talked Fermi was looking at a bright star, and by moving his head so the wires would hide it from sight, he was observing the scintillation.

We agreed that the affair would ultimately lead to a beatification of Oppenheimer; he would become a great martyr and his accusers would be damned. Both Fermi and von Neumann were, in the hearings, fully on Oppenheimer's side and defended him against the accusations, though neither was a great personal friend or special admirer of his. Fermi was not greatly impressed by his physics and had some reservations about his political leanings. He felt however that Oppenheimer had been treated very shabbily. We also discussed the attitude of Edward Teller, and I asked Fermi how he viewed the future. Suddenly he looked at me and said: "I don't know, I'll look at it from up there," pointing at the sky. Did he have some premonition that he was dangerously sick? If so, he never admitted it in so many words nor did he look it. But this struck me as a bolt from the blue, especially since he repeated it once more as we discussed the foundations of physics, the mysteries of particles, the behavior of mesons, and his changing interests from nuclear structure to the supposedly more fundamental parts of the physics of particles. Again he said: "I'll know from up there." The next day we separated, the Fermis driving east to Grenoble and Les Houches, and we, Claire, and Françoise's brother south, to spend a vacation in La Napoule, near Cannes on the French Riviera.

When we returned to the United States at the end of the summer, the news was that Fermi was very ill, that an exploratory operation had been performed immediately upon his return to Chicago and that a generalized cancer of the esophagus and stomach had been found. Some of his friends thought the cancer might have been caused by his early work with radioactive materials at a time when precautions were not very carefully observed. I wondered then whether a habit of his which I had noticed of occasionally swallowing hard, and which I thought was a deliberate form of self-control, might have been connected all along with a physical difficulty.

His illness progressed rapidly. I went to Chicago to visit him. In the hospital I found him sitting up in bed with tubes in the veins of his arms. But he could talk. Seeing me, he smiled as I came in and said: "Stan, things are coming to an end." It is impossible for me to describe how shattering it was to hear this sentence. I tried to keep composed, made a feeble attempt at a joke, then for about an hour we talked about many subjects, and all along he spoke with serenity, and under the circumstances really a superhuman calm. He mentioned that Teller had visited him the previous day, and joked that he had "tried to save his soul." Normally it is the priest who wants to save the soul of the dying man; Fermi put it the other way round, alluding to the public hullabaloo about Teller and the H-bomb. Perhaps their conversation had an effect, for shortly after Fermi died Teller published an article entitled "The Work of Many People," toning down the assertions of Shepley and Blair. During my visit to Fermi Laura dropped in and I was amazed at the ordinary nature of their conversation about some household appliance.

We talked on and I remember his saying that he believed he had already done about two-thirds of his life's work, no matter how long he might have lived. He added that he regretted a little not having become more involved in public affairs. It was very strange to hear him evaluating his own activity—from the outside, as it were. Again I felt that he achieved this superobjectivity through sheer will power.

Somehow the conversation turned to the progress of medicine. He said, "Well Stan, you know, my chance of living through this is perhaps not zero but it is less than one in a hundred." I looked at him questioningly and he continued, "I believe that in twenty years or so a chemical cure for cancer will be found. Now I have only two or three months, and assuming uniform probabilities, the ratio of these times is one hundred to one." This was his characteristic way of trying to be quantitative, even in situations where it is not possible. Then half seriously I raised the question whether in a thousand years so much progress will be made that it may be possible to reconstruct people who had lived earlier by tracing the genes of the descendants, collecting all the characteristics that make up a person and reconstructing them physically. Fermi agreed, but he added: "How about the memory? How will they put back in the brain all the memories which are the makeup of any given individual?" This discussion now seems rather unreal and even weird, and it was partly my fault to have put us on such a subject, but at the time it came quite naturally from his super-detachment about himself and death. I paid him one more visit, this time with Metropolis; when we came out of his room I was moved to tears. Only Plato's account of the death of Socrates could apply to the scene, and paraphrasing some of the words of Krito I told Nick, "That now was the death of one of the wisest men known."

Fermi died shortly after. A short time later I was passing through Chicago again and called on Laura. I gave the address to the driver and added that this was the house of the widow of the famous Italian scientist who had just died. The driver, who happened to be Italian and who had read about it in the papers, absolutely refused to let me pay the fare. Only when I told him he could give the money to charity did he take it.

Just after Johnny was offered the post of AEC Commissioner and before he accepted and became one in 1954 we had a long conversation. He had profound reservations about his acceptance because of the ramifications of the Oppenheimer Affair. He knew that the majority of scientists did not like Admiral Strauss's actions and did not share the extreme views of Teller. Some of the more liberal members of the scientific community did not like Johnny's pragmatic and rather pro-military views nor did they appreciate his association with the atomic energy work in general and with Los Alamos in particular, especially his contributions to the work on the A and H bombs. He recognized this feeling even among some of his Princeton associates, and he was afraid that it would become stronger when he joined the Atomic Energy Commission. This despite the fact that in the Oppenheimer Affair, even though he did not especially like Oppenheimer personally, he defended him with great objectivity and gave very correct, courageous, and intelligent testimony.

The decision to join the AEC had caused Johnny many sleepless nights, he said, and in a two-hour visit to Frijoles Canyon one afternoon he bared his doubts and asked me how I felt about it. He joked, "I'll become a commissionnaire." (In French the term is used to mean errand boy.) But he was flattered and proud that although foreign born he would be entrusted with a high governmental position of great potential influence in directing large areas of technology and science. He knew this could be an activity of great national importance. Indeed, with his supreme intelligence, he could have done an enormous amount of good in seeing what was valuable in certain programs and in initiating new ones. As a friend of his and having pressed him to accept the offer, Strauss would be obligated to support his views and ideas. Besides, Johnny had a bit of the Teutonic trait of being easily impressed by officialdom. At any rate, he was torn between two poles: a feeling of pride with the hope of doing something good and useful and the fear of becoming associated in the minds of his colleagues with a small minority of the scientific community and of career-oriented persons. Acceptance required taking a leave of absence from the Institute and some financial sacrifice as well. I do not know the details of the promises that Strauss may have made or the pressures he may have exerted.

I wondered later whether this decision and the anguish and nervous tension it caused him had perhaps predisposed him to the onset of the fatal illness that came not long after. Obviously taking this step did have some physical impact, for he looked wan and showed the effects of stress. Besides, there was a lot more physical work involved. I don't think that ever before he had worked from eight in the morning to five in the afternoon in the same place and with several meetings every day. No matter how hard he had worked before it had been on his own time and choosing. The first indication that something was very wrong with his health appeared some time after he became a commissioner.

As I had known him over the years he always seemed in good health. He only had infrequent colds, slept well, worked hard, could eat and drink liberally without showing any effects. I don't think he was hypochondriac. On the contrary, except for an occasional cold or toothache, he was very little preoccupied by his own physical state, although he once showed me some correspondence he had had with Dr. Janos Plesch about kidney function. Once on one of our many walks in Frijoles Canyon he remarked in passing a tree throttled by a vine, how horrible it must feel to be surrounded and trapped and unable to get away. This remark came back to my mind later when he became paralyzed.

I got wind of some vague rumors that he was ill. I asked Teller about it, but he gave me evasive answers and said something I could not interpret. I telephoned the house in Georgetown, and Klari told me a noncommittal little story. I could not help but suspect that something was very wrong, and later found out that Johnny had given specific orders that I should not be told that he had developed cancer. One day sitting in his office he had been seized by a violent pain in the shoulder, so strong that he almost fainted. This pain disappeared, but he went to Massachusetts General Hospital in Boston where a small cancerous growth was removed from his clavicle, probably already a secondary growth. He soon recovered from this surgical intervention and came to Los Alamos for what was to be his last visit there. I still had not been told what was wrong with him.

He came to our house and I noticed that he limped slightly. He seemed obviously preoccupied and perturbed. There was a sadness in him and he frequently seemed to look around, as if, it occurred to me later, he might have been thinking that this was perhaps his last visit and he wanted to remember the scenery, the mountains, the places he knew so well and where he had so often had interesting and pleasant times. Yet at the same time he joked about his presence in Los Alamos as a commissioner. Now he was there not only to think about scientific matters but about very prosaic administrative ones. And Rabi who was in town also chimed in that it was no longer a scientific visit but an inspection tour. Before he left, Françoise showed him a recent snapshot of Claire on her bicycle. He asked if he could take the photo with him. He walked back to the Lodge through the garden, and watching him through the window I definitely had the feeling that somber and melancholy thoughts were in his head.

A few weeks later on one of my visits to Washington Johnny took me to lunch. During the meal he told me that doctors had discovered he had cancer and he described what kind. This was a tremendous shock for me. I told him my suspicions that something was wrong and that for some reason I had wondered if it could be diabetes or his heart. I turned away so as not to show how upset I was, but he noticed it anyway and started telling a joke about a woman in Budapest whose maid had fallen ill. She sent for a doctor who told her that her maid had syphilis. "Thank God," the woman said, "I was afraid it was measles and she would infect the children." During this dramatic lunch he still showed great strength of will and no signs of knuckling under. I was shattered and wondered whether he would ever recover.

On my next trip I visited Johnny at home. The Georgetown house he had rented was very different from his Princeton one. It was small, very seventeenth-century Dutch with a black-and-white-tiled vestibule as in some of Vermeer's paintings. He was still working at the AEC but walked with increasing difficulty and soon had to take to a wheelchair. Friends and even doctors wondered whether some of it was psychosomatic. It was never clear what kind of cancer he actually had. I never learned the whole story; I don't think many people knew it. Klari would never say much about it. I was told it started in the prostate and ultimately metastasized so that he became partly paralyzed.

During his illness Johnny did not talk to me about his important work on the ICBM Committee; only later I learned that he was Chairman and that it was called the von Neumann Committee. As he grew increasingly ill, some of the meetings were held in his house and later at the Woodner Hotel to which the von Neumanns had moved so as to be closer to Walter Reed Hospital, where he was undergoing treatment. To the end he maintained this complete discretion. Even though I was perhaps one of his closest friends he never broached to me classified or military subjects in which I was not involved. Our usual conversations were either about mathematics or about his new interest in a theory of automata. These conversations had started in a sporadic and superficial way before the war at a time when such subjects hardly existed. After the war and before his illness we held many discussions on these problems. I proposed to him some of my own ideas about automata consisting of cells in a crystal-like arrangement. This model is described in the book edited by Arthur Burks, *Cellular Automata*, and in Burks's own book on the theory of automata. At that time it was believed that there were 10^{10} neurons in the small space of the human skull and that from some neurons there issued some hundreds and perhaps in the center region a thousand connections with other neurons. We used to marvel at the complication of the organization of the brain. Now I understand that it has been found that there are thousands of connections from each neuron to others and in some areas fifty thousand and even more. And each neuron, which at that time was believed to be just a rather simple "flipflop" "yes or no" machine, is now believed to be a complicated organ with many more functions. In the space of fifteen years, since von Neumann's death, the facts have become amplified; the whole structure is even more amazing, more incredible than it seemed at the time. Johnny did not live to see the developments following Crick and Watson's work on the structure of the DNA chains in the nucleus of cells and the code which they contain.

It is evident that Johnny's ideas on a future theory of automata and organisms had roots that went back in time, but his more concrete ideas developed after his involvement with electronic machines. I think that one of his motives for pressing for the development of electronic computers was his fascination with the working of the nervous system and the organization of the brain itself. After his death some of his collaborators collected his writings on the outlines of the theory of automata. Published posthumously, his book on the brain had merely the barest sketches of what he planned to think about. He died so prematurely, seeing the promised land but hardly entering it. The great developments in molecular biology really came too late for him to learn much about it and to enter a field which I know fascinated him.

Another source of stimulation came from his interest in the theory of games. This was initially perhaps an independent curiosity, but in my opinion, a general theory of contests, fights for survival, and evolution will furnish in the near future a whole class of new mathematical problems and new patterns of thought concerning the schemata of the development of biological processes through what is now called evolutionary and "survival of the fittest" competition. In that area one of his major undertakings was the elaboration and creation of new models of probabilistic theory of games, in particular the study of the rules of coalitions. He developed these ideas with Oskar Morgenstern, a Princeton economist, in a monumental book entitled *Theory of Games and Economic Behavior.*

In the space of fifteen years since von Neumann's death the new facts discovered have become more perplexing, the whole structure is even more amazing, more incredible than it seemed at the time. It will go on increasing as our understanding of anatomy and physiology improves and will lead to new fields of mathematical research.

This process of increasing complexity in science is going on with no sign as yet of slowing down. Whether it will continue indefinitely or regress is a big question. It is a part of the problem of infinity versus the finiteness of the world.

In the last months of his life, Johnny was hospitalized at Walter Reed Hospital. He occupied a very large suite reserved for high government officials. In the fall of 1956 we were living in Cambridge again, and I was a visiting professor at MIT on another leave from Los Alamos. I managed to travel to Washington and visit him a few times. On one of these visits once again we had a discussion about age. He wondered how much more original and creative work he could still do if he lived. I tried to encourage him by telling him that he still could do at least half as much again.

Curiously, three years earlier while visiting Fermi in his hospital in Chicago, our conversation had also turned to the same topic; Fermi had said calmly that he considered he had already done most of his work. What a difference in outlook, or at least in the way these two great men expressed or suppressed their feelings.

On the same visit I went by mistake to the opposite corner of the hospital but on the same floor, and walked into an antechamber where two military men were sitting. They looked at me in surprise and questioningly. I said I was there to visit a friend and their look turned incredulous. When I added, "Dr. von Neumann," they smiled and directed me to the proper rooms. I had entered the Presidential Suite where President Eisenhower at that moment was hospitalized after his heart attack. I told this to Johnny when I regained his room. He enjoyed this. It amused him to be in a location symmetrically opposite to that of the President of the United States.

Some months before, Admiral Strauss had a conversation with me about what Johnny's life could be should he recover sufficiently to leave the hospital but not sufficiently to rejoin the Commission. The idea was to cheer him up with new surroundings and perhaps provide him with a perspective on things other than governmental work. Strauss, though not believing that a full recovery would come to pass, was instrumental in obtaining for him an offer of a special professorship at UCLA. This prospect diverted and cheered Johnny somewhat.

He never complained about pain, but the change in his attitude, his utterances, his relations with Klari, in fact his whole mood at the end of his life were heartbreaking. At

(243)

one point he became a strict Catholic. A Benedictine monk visited and talked to him. Later he asked for a Jesuit. It was obvious that there was a great gap between what he would discuss verbally and logically with others, and what his inner thoughts and worries about himself were. It was visible on his face. Johnny used to be completely agnostic even though he sometimes expressed his feelings of wonder and mystery. Once in my presence when Klari chided him for his great self-confidence and pride in his intellectual achievements, he replied that on the contrary he was full of admiration for the wonders of nature and the evolution of the brain, compared to which all we do is puny and insignificant.

By then he was very, very ill. I would sit with him and try to distract him. There was still some scientific curiosity in him; his memory still seemed to work sporadically, and on occasion almost uncannily well. I will never forget the scene a few days before he died. I was reading to him in Greek from his worn copy of Thucydides a story he liked especially about the Athenians' attack on Melos, and also the speech of Pericles. He remembered enough to correct an occasional mistake or mispronunciation on my part.

Johnny died in Walter Reed hospital, February 8, 1957. He was buried in Princeton in a brief Catholic service with a short eulogy by Admiral Strauss. After the funeral there was a small gathering at his house. Several mathematicians were there, among them his old friend James Alexander, himself recovered from an illness not unlike the one I had had in Los Angeles. Also present were Atle Selberg, the number theorist, and Lewis Del Sasso, an engineer who had worked at building the MANIAC, and Mrs. Gorman, his long-time secretary at the Institute. After his death, Françoise went to Washington to spend a few days with Klari, taking Claire along. The presence of a child she was fond of helped momentarily to take Klari's mind off the long grueling months that preceded Johnny's death.

Von Neumann was remarkably universal. I have known

wonderful mathematicians who were severely limited in their curiosity about other sciences but he was not.

Von Neumann's reputation and fame as a mathematician and as a scientist have grown steadily since his death. More than his direct influence on mathematical research, the breadth of his interests and of his scientific undertakings, his personality and his fantastic brain are becoming almost legendary. True, in his lifetime he had already achieved an enormous reputation, and all the honors the mathematical world can give. But he had his detractors. He was not entirely what one might call a mathematicians' mathematician. Purists objected to his interests outside of pure mathematics when, very early, he leaned towards applications of mathematics or when he wrote, as a young man, about problems of quantum theory.

As for myself, I was never greatly impressed by his work on Hilbert space or on continuous geometry. This is a question of taste, and when I was more of a purist myself I made good-natured fun of certain of his involvements in applications. I told him once, "When it comes to the applications of mathematics to dentistry, maybe you'll stop."

But there was nothing small about his interests, and his exquisite sense of humor prevented him from going off on tangents from the main edifice of mathematics. He was unique in this respect. Unique, too, were his overall intelligence, breadth of interest, and absolute feeling for the difference between the momentary technical work and the great lines of the life of the mathematical tree itself and its role in human thought.

Now Banach, Fermi, von Neumann were dead—the three great men whose intellects had impressed me the most. These were sad times indeed.

PART IV

The Past
Fifteen Years

CHAPTER 13

Government Science

1957–1967

I T is more difficult to write about recent events; the perspective is poorer, the separation of the characteristic or the important from the fortuitous is more difficult. My story of the past fifteen years or so will therefore be contracted and will concern activities and people even more arbitrarily chosen than the reminiscences and reflections of the earlier chapters.

Returning to Los Alamos in 1957 after a year's leave of absence at MIT, I was asked by Bradbury to accept one of the two newly created positions of research advisor to the director of the laboratory. The other advisor was to be John Manley, a physicist who had held important administrative posts at Los Alamos during the war and wanted to return to New Mexico after a long absence as professor at the University of Washington in Seattle. Administratively, research advisors were to be on the same level as division leaders, and their duties were to oversee the research activities throughout the laboratory, in the various divisions: theoretical, physics, chemistry and metallurgy, weapons, health, "Rover" (nuclear rocket), and others. Together we tried to influence the various programs of the lab. This was an arduous and many-sided task, and talking with many people about their research activities enabled me to broaden my own interests. I held this position until 1967 when I retired from Los Alamos and joined the mathematics department at the University of Colorado in Boulder. In laboratory jargon, Manley and I were respectively known as RAJM and RASU.

In my administrative role in Los Alamos and later as chairman of the mathematics department in Boulder, I came to understand and appreciate better and also commiserate with friends and acquaintances who had become fully occupied by administrative duties. In my younger years I had had the usual skeptical attitude towards most chairmen of departments, deans, presidents, directors, and the like. There had been exceptions, of course. One of these was J. Carson Mark, leader of the theoretical division in Los Alamos since the middle forties. Mark is a Canadian mathematician who came to Los Alamos towards the end of the war with the British Mission. He became an American citizen some years later. He bore the brunt of the difficulties with Teller with remarkable calm and objectivity; he is one of the few mathematicians I know who have an understanding for the problems of physics and associated technology in a broad sense. His direction of the theoretical division was an example of intelligent management of a scientific group without exercising undue pressures for programmatic work. He was able to encourage free scientific pursuits in areas which were only indirectly related to the tasks of the laboratory, and he supported theoretical physics and applied mathematics in the best sense. (Incidentally, he was also a regular participant of our poker sessions. From 1945 until now I cannot recall a single occasion when he refused to come or missed one of our games. These by now have considerably slowed in frequency. From weekly they became monthly and now occur only occasionally, mainly when I happen to be visiting Los Alamos.)

After the war it had become clear that science and technology had become so crucial for national affairs that the governments of the Western world had to devote enormous amounts of time and huge budgets to them. Famous scientists were called upon to enter the inner circles of government to help direct their countries' scientific activities, not only for the arms race but for technological advancement. Churchill had had Lord Cherwell; De Gaulle, Francis Perrin; America, her Scientific Advisory Committees. Beginning with Bush and Conant, Oppenheimer, von Neumann, and many others became government "sages." Government science peaked and committees proliferated under the Eisenhower, Kennedy, and Johnson administrations, and even I found myself called upon. Until then I had always resisted being drawn into any kind of organizational position; for years I could claim that my only administrative job had been on the Wine Tasting Committee of the Society of Fellows back in my early Harvard days.

A few years before Johnny's death and increasingly so after, as a result of my work on the hydrogen bomb, I became drawn into a maze of involvements. These had to do precisely with government science and with work as a member of various Space and Air Force committees. Also, in some circles I became regarded as Teller's opponent, and I suspect I was consulted as a sort of counterweight. Some of these political activities included my stand on the Test Ban Treaty and testimony in Washington on that subject. The cartoonist Herblock drew in the *Washington Post* a picture of the respective positions of Teller and me in which I fortunately appeared as the "good guy."

Since I never have kept notes or diaries of any kind, I may not always be entirely correct about the chronology of events or how things and people were connected during these busy years of scientifico-technological activities.

Washington committees, I soon noticed, were often very

envious of new ideas, with their members exhibiting the well-known "not invented here" syndrome of rejecting proposals or ideas only because of the vested interests of committee members. This feeling was a greater obstacle to the development of new projects than the concern about their cost and the amounts of money they would require. Decisions also seemed sometimes dictated, not so much by objective evaluation as by the usual academic rivalries and envy of scientific fame. Had I not been already quite old and cynical this would have made me leave governmental science altogether. I remembered how in Los Alamos Johnny remarked on several occasions that it was not easy to introduce new things; one had to persuade every janitor, he said. But once something was accepted it became a sort of bible and it was equally hard or even impossible to change it or get rid of it. This national situation has become even worse now, partly because of the recent spreading skepticism regarding the value of science and its benefits, and a kind of general passivity so removed from the traditional traits of American enterprise, energy, and spirit of cooperation.

The idea of nuclear propulsion of space vehicles was born as soon as nuclear energy became a reality. It was an obvious thought to try to use its more powerful concentration of energy to propel vehicles with a very large payload for ambitious space voyages of exploration or even for excursions to the moon. I think Feynman was the first in Los Alamos during the war to talk about using an atomic reactor which would heat hydrogen and expel the gas at high velocity. A simple calculation shows that this would be more efficient than expelling the products of chemical reactions.

I became involved with two such projects, one in an advisory capacity, with the other more directly. The first was Project Rover, a nuclear-reactor rocket which was being designed in Los Alamos already quite a few years before the Russian Sputnik, but with very limited funds. The second was a space vehicle, later named Orion. Around 1955 Everett and I wrote a paper about a space vehicle propelled by successive explosions of small nuclear charges. The idea has even been patented by the AEC in our names. This method could be much more powerful than Rover and is a very ambitious but efficient way to undertake space explorations with a vehicle able to travel at high speeds with high payloads and an extremely good ratio of payload to total initial weight. The spaceship could transport hundreds or thousands of people. When Kistiakowski was President Eisenhower's Scientific Advisor I informed him about such possibilities, but his reception of it was not enthusiastic. But more about Orion later.

Soon after John Kennedy's election in November 1960, I received a telephone call from Jerry Wiesner from Cambridge. I had met Wiesner the year I was at MIT as a visiting professor; we had seen each other several times and had had good conversations about science projects, national programs, education, and so on. We had talked also about the Teller business. Wiesner was wary of the Edwardian brand of politics. I was not too surprised when I received this call. It must have lasted more than half an hour and Jerry informed me that President Kennedy had appointed him chairman of a task force on science and technology. He asked what my ideas were about the nationally important scientific or technological projects the President should know about and consider for the country. "How about going to the moon?" I asked. I imagine dozens of other people had made the same kind of suggestion. In his inaugural address Kennedy proposed a national project to put a man on the moon. My involvement with the space effort began in earnest with that conversation. I became consultant to Wiesner's Scientific Advisory Committee and visited Washington frequently.

Immediately after the war, Clinton P. Anderson, Senator from New Mexico, a former member of President Truman's cabinet, became one of the most interested, knowledgeable, influential, and effective proponents of the uses of nuclear energy. He was instrumental in helping the Los Alamos Laboratory and the associated big installation in Albuquerque, the Sandia Laboratory.

I became acquainted with him during one of his early visits to Los Alamos, enjoying his confidence and—it seems to me—his trust in and reliance on my opinions, not only in the area of nuclear energy but also in the field of space activities. Several times he invited me to testify before Congress on specific space matters, such as the organization of NASA and whether it should be part of the military establishment or an independent organization.

When it was decided to do something in earnest about Project Rover, Wiesner named a Presidential Committee to look into the matter. I was one of its members. Among some members who were chemists I noticed a degree of skepticism about its worth and feasibility, again motivated in my opinion by their apprehension that it might compete with the already existing chemical rocket propulsion systems which were being developed. Some of the discussions reminded me of the big debates at the beginning of the century between lighter-than-air and heavier-than-air advocates, or even of the earlier competition of steamships versus sailing vessels. And indeed the committee wrote a report which by faint praise, essentially condemned Project Rover to a de facto death by proposing to make it a purely theoretical study without funds for experimental work or any investment in construction. The physicist Bernd Matthias was the only member of the committee who joined me in writing a dissenting opinion.

Senator Anderson was chairman of the congressional Space Committee. He knew my position on Rover. With his feeling for the psychological and political motivations of committees, and his vital interest in the new technology and its importance for the nation, he took me one day to the office of the then Vice-President Johnson. Together we walked to a nearby building to see Wiesner. Since Jerry and

I were friends, it embarrassed me a little to be present at a meeting in which Johnson and Anderson were pressing him hard to change the attitudes of the Scientific Advisory Committee on questions of nuclear propulsion. They supported my views in opposition to his. Ultimately the minority opinion prevailed, and the Rover Project was saved. Funds were allotted for Los Alamos work and over the years it became an extremely successful venture. Unfortunately it was stopped again later by economies in the space program.

I was also invited to join an Air Force Committee on a similar subject: general problems of plans for space and the Air Force's role in it. The Committee was chaired by Trevor Gardner, a former assistant secretary of the Air Force during Eisenhower's presidency. Gardner was a very interesting person of whom I became very fond. His vigorous and lusty personality, his great energy, the wide scope of his imagination appealed to me very much. I found him very congenial.

The committee originally comprised a number of persons important in science and technology. The only other mathematician, Mark Kac, was present at a few sessions. Among the "big shot" members, I remember Harold Brown, director of the Livermore Laboratory, later Secretary of the Air Force, Charlie Townes, who received the Nobel Prize for the invention of masers, General Bernard Shriever, a frequent visitor. Vince Ford, an Air Force colonel who had been Johnny's aide on the von Neumann ICBM Committee, was now Gardner's assistant. He organized the meetings of a working subcommittee which met in Los Alamos. These meetings involved many people from the newly born aerospace industry. Sometimes we met in Los Angeles where the headquarters of the ballistics division of the Air Force under General Shriever were located. At other times we met in Washington where General Shriever, Gardner, Ford and I discussed among ourselves during restaurant lunches how to plan the exploration of space and more generally the problems of space study for the Air Force.

Early at one of these meetings, somebody from industry presented plans for retrieving rocket engines, which would save money by making them reusable. The real problem, as I and some others saw it, was to do something important in putting up the satellites and do it quickly rather than to start by saving money. Also it seemed to me that boosters, namely the engines, were a small part of the overall cost, and that it would be awkward, to say the least, to start by reusing second-hand engines, perhaps damaged. When the proponents droned on about their ideas, showing their plots and graphs, I whispered to Gardner, "This sounds to me like a proposal to use the same condom twice." He burst out laughing and sent the remark around the table in repeated whispers. Perhaps this joke saved the United States some millions of dollars in expenditure for what would have been pointless and impractical work at the time.

Orion was also discussed by the Gardner Committee. At my suggestion Ted Taylor became the executive director of a group working on it. Starting in 1957, Taylor developed the Orion idea as a reaction to the Russian Sputnik, as he said. He assembled an impressive group of bright young men at the General Atomic Laboratory in La Jolla, Calif. The physicist Freeman Dyson became very interested and enthusiastic and took a leave of absence from the Princeton Institute to work for a year with Taylor. A few years later he wrote an eloquent article describing the project and how it was put on a shelf. It appeared under the title "La Vie et Mort d'Orion" in the Paris paper *Le Monde*.

Somehow the Gardner Committee report got lost at a high level in the Washington maze. Wiesner disagreed with Gardner about the role of the Air Force in space. Somebody in Washington managed to bury the report, and I don't think President Kennedy ever saw it. The whole thing is still a mystery to me. After the Gardner Committee finished its work it was succeeded by another one called the Twining Committee. Its members included some hawkish types like

Teller and Dave Griggs. General Doolittle of Tokyo air raid fame was also a member.

I became connected with Trevor Gardner later in a more private capacity. He asked me to join the scientific advisory board of the Hycon Corporation in California which he headed. The company manufactured highly secret military equipment including special cameras. Fowler, Lauritsen, Al Hill, a physicist from MIT, and Jesse Greenstein the Palomar astronomer, were the other members of the board. I learned that Wiesner and his group in Hycon East and Gardner in California had had some serious disagreements about financial problems concerning the corporation, and apparently Wiesner and Gardner were barely on speaking terms.

To some Gardner was a controversial person because of his quick temper and his strong opinions. He had great political ambitions (he would have liked to become Secretary of Defense), but he was at cross purposes with some members of the Kennedy Administration. He died of a heart attack shortly before Kennedy's assassination. It was Gardner who had established the von Neumann ICBM Committee. This had been of immense importance for the U.S. space effort; I think it really got it off the ground. The military and national importance of this and other Gardner initiatives can hardly be exaggerated.

At the same time, I was continuing my own work. After Fermi's death Pasta and I decided to continue exploratory heuristic experimental work on electronic computers in mathematical and physical problems. We felt that the combination of classical mechanics and astronomy problems lent itself to two kinds of studies: one, the behavior of large numbers of particles—call them stars—in a cluster or galaxy; the other, the history of a single mass of gas as it developed from initial conditions by contraction at first, perhaps giving rise to a double or multiple star, then generating more and more nuclear reactions, exhausting its nuclear material, and finally perhaps collapsing. Many calculations have been done since on this latter problem over the last twenty years, changing the whole understanding of astrophysics and of the development of the universe as far as individual stars are concerned.

The problem of clusters of stars was I think the first study of this nature using computers. We took a great number of mass points representing stars in a cluster. The idea was to see what would happen in the long-range time scale of thousands of years to the spherical-looking cluster whose initial conditions imitated the actual motions of such stars. This was a really pioneering calculation showing that this sort of investigation was possible. It gave very curious and unexpected observations in classical mechanics, formation of subgroups, and contractions. We made a film of these motions on an accelerated scale which showed these interesting phenomena. This work gave rise to other studies of this sort at Berkeley, in France, and elsewhere.

Another problem which I attacked but which is still not solved is an attempt to see what will happen when a mass of gas of very large dimensions, say of the whole solar system, at very low density and having initially a mild amount of turbulence starts to contract. How would it contract, and how would it finally form a star? What is interesting and the actual purpose of the problem was to see whether and how often it would form a double, triple, or multiple star. The reason for this curiosity is that many stars, in our neighborhood at least, are double. According to recent studies, at least one star out of three is multiple. It would be nice to see by brute-force calculations how a contraction of an irregularly shaped mass of gas develops. Beyond all this is the problem of the formation and development of galaxies—that is, the assemblies of billions of stars. On this, too, astrophysicists have accomplished much with the aid of computers.

While such astrophysical calculations were going on, I began in an amateurish way to work on some questions of biology. After reading about the new discoveries in molecular biology which were coming fast, I became curious about a conceptual role which mathematical ideas could play in biology. If I may paraphrase one of President Kennedy's famous statements, I was interested in "not what mathematics can do for biology but what biology can do for mathematics." I believe that new mathematical schemata, new systems of axioms, certainly new systems of mathematical structures will be suggested by the study of the living world. Its combinatorial arrangements may lead us in the future to a logic and mathematics of a different nature from what we know now. The reader is referred to one of my papers on mathematical biology. Too technical to be included here, it is listed in the bibliography at the end of this volume.

My interest in biology took a more tangible form when I engaged in discussions with James Tuck, and we talked to the biologists in the laboratory. Los Alamos had always had a division for the studies of biological effects of radiation. Radioactive damage was, of course, one of the first things to worry about from the beginning of the nuclear age. With Tuck, and Gordon Gould and Donald Peterson from the Health Division, we organized a seminar devoted to current problems of cellular biology and the new results in molecular biology. I really learned a lot about the elementary facts of biology there, the role of cells, their structure, and so on. The seminars, which had about twenty participants, have had important consequences, although they lasted only two years. Two of the participants, Los Alamos physicists Walter Goad and George Bell, both extremely brilliant and talented, and among the best young brains in these fields in the country, are doing a lot of biology research now. Goad is working in the field of biological mathematics, while Bell has some new ideas on immunology. Ted Puck from Colorado visited the seminar and gave some lectures.

I met Puck shortly after the war and found him to be full of new ideas, suggesting interesting experiments and methods for the study of the behavior of cells and problems in molecular biology in general. I think it was Ted Puck's group which first succeeded in keeping mammalian cells alive and even multiplying in vitro. I always look forward to discussions with him; it was he who arranged for me to give seminars for the faculty and young researchers in the biophysics department and even succeeded in having me appointed a member of the professorial staff at the University of Colorado's Medical School. I told him that, being a beginner and a layman in this field, I might be arrested for impersonating a doctor.

Almost every month there are fascinating new facts discovered in biology. It is now widely recognized that the discoveries of Crick and Watson have opened up a new era in the psychological attitudes in biology as well. Years ago at Harvard, when I talked to biologists and tried to ask about or propose even a mildly general statement, there would always be the retort: "It isn't so because there is an exception in such and such an insect" or "such and such a fish is different." There was a general distrust or at least a hesitation to formulate anything of even a slightly general nature. This attitude has drastically changed since the discovery of the role of DNA and the mechanism of replication of the cell and of the code which seem so universal.

During all these years I did not live continuously in Los Alamos. I spent periods of time as a visiting professor at Harvard, MIT, the University of California in La Jolla, the University of Colorado, plus innumerable visits to various universities, scientific meetings, and government or industrial laboratories, where I gave lectures and consultations. These latter were called business trips. If one adds our almost yearly vacations in Europe since 1950 (mainly in France where Françoise still has relatives and I have many scientific friends), it seems to me that about twenty-five percent of my time was spent away from Los Alamos.

It was in those periods that my friendship with Victor Weisskopf developed. I had met him in Los Alamos during the war when he was Bethe's alternate as leader of the theoretical division. He left at the war's end to become professor at MIT and our relationship deepened during my visits to Cambridge, Harvard and MIT.

Viki, as he is universally called, is a theoretical physicist. He made a name for himself as a young man with his important work on problems of radiation in quantum theory. He was for a time assistant to Pauli and also worked in Copenhagen at the famous Niels Bohr Institute. Viki was born in Vienna, a fact I take note of because he exhibits the best side of the Viennese temperament. This is contained in the following saying: In post-World-War-I Berlin people used to say, "The situation is desperate but not hopeless"; in Vienna they said, "The situation is hopeless but not serious." This certain insouciance combined with the highest intelligence has enabled Viki to navigate not only through the usual difficulties of administrative and academic affairs—he has been, among other things, director-general of CERN (the European Center for Nuclear Research) near Geneva and chairman of the large physics department at MIT—but also in the more abstract realm of the intellectual and scientific difficulties of theoretical physics. I would say his intellectual stability is based on a real knowledge of and feeling for the spirit of the history of physics. This he has achieved through perspective and comprehension, sifting and evaluation of the quickly changing scene in the physical theories which concern the very foundations of this science. I should add here for the benefit of the reader who is not a professional physicist that the last thirty years or so have been a period of kaleidoscopically changing explanations of the increasingly strange world of elementary particles and of fields of force. A number of extremely talented theorists vie with each other in learned and clever attempts to explain and order the constant flow of experimental results which, or so it seems to me, almost perversely cast doubts about the just completed theoretical formulations. Through all this turmoil in the overly mathematical theoretical physics research, constant good progress has been made, but it takes a person like Viki (and really there are no more such than one can count on the fingers of one hand) to stabilize this flow and extract the gist of the new elaborations of the ideas of quantum theory and to be able to explain and describe it both to the physicists themselves and to the more general public.

His semipopular books on physics are uniquely interesting and successful in presenting his philosophy and the human side of the story. He was always and still is immensely concerned with the problems of man and world affairs. As a person he is affable and kind, gets along with everybody, and he loves to tell stories, and sometimes our exchanges of Jewish jokes can last an hour.

During his tenure at CERN where he still visits every summer and consults, the Weisskopfs built a modest summer house on the French side of the border in a small Jura village overlooking the Lake of Geneva. It is twenty minutes' drive from CERN and they spend most of the summer there. On our own European trips of the last several years we have almost always included a brief stay with the Weisskopfs in their house in Vesancy. This village is just a few kilometers away from another one known as Ferney-Voltaire because Voltaire lived there for many years. In like manner I have dubbed Vesancy, Vesancy-Weisskopf. Viki likes that and it fits for he has become quite a personage in the village. He is known as Monsieur le Directeur, and the farmers tip their hats when they see his tall, lanky silhouette walking carefully across their fields.

In 1960 my book, *Unsolved Problems of Mathematics*, was published. Many years ago Françoise asked Steinhaus what it was that made me what people seemed to consider a fairly good mathematician. According to her, Steinhaus replied: "C'est l'homme du monde qui pose le mieux les problèmes." Apparently my reputation, such as it is, is founded on my ability to pose problems and to ask the right kind of questions. This book presents my own unsolved problems. As a young man I liked the motto in front of

George Cantor's thesis which is a Latin quotation: "In re mathematica ars proponendi quaestionem pluris facienda est quam solvendi."

Shortly after 1960 the book was translated into Russian. There is no copyright agreement between Russia and the West, and the Russians pay no royalties, but some Western authors discovered when they were in the Soviet Union that they could obtain some payment for the translations of their work. Hans Bethe and Bob Richtmyer successfully received compensation. So when I attended an International Mathematics Congress in Moscow in 1966 I remember that I could try too. The Russian language being close to Polish, I went to the publishing house to talk about this matter in my imitation of Russian. At the publishing house, which looked the same as everywhere else—girls typing and masses of files and papers—an elderly gentleman seemed to understand my request and asked me how I knew to come and see them. I gave him the names of my friends. He went to a back room, then returned. The reader should know that Russians do not pronounce H as in English. They say G. For instance Hitler is pronounced Gitler, Hamlet Gamlet, Hilbert Gilbert. The gentleman said to me in Russian with an engaging smile: "Come back tomorrow please with your passport, and we will give you"—I seemed to hear—"your gonorrhea." Of course he had really said "gonorar" for "honorar" (honorarium or royalties). I wanted to say, "No, thank you," but I understood what he meant. The next day when I returned he handed me an envelope which contained three hundred rubles in cash. One is not allowed to export rubles from Russia, so after I had bought some souvenirs, amber, fur hats, books and the like, I still had one hundred rubles left. I had to put them in a postal savings account which in Russia pays one or two percent interest. This makes me a Soviet Union capitalist.

Sometime in the early 1960s I met Gian-Carlo Rota, a mathematician who is almost a quarter-century younger than I and definitely representative of the next generation. Or maybe even *several* generations later, for academically in mathematics the generation gap may exist already between lecturers and their students where chronologically the difference is only a few years. Our relationship is not built on our age difference. Rota claims that he is greatly influenced by me. So I coined the expression "influencer and influencee." Rota is one of my best influencees. Banach, for example, I consider as an influencer.

From the start I was impressed by Rota's feeling for several different mathematical fields and his opinions in many areas of research where he exhibits both erudition and common sense. It is increasingly rare now—in fact, it has been for the last twenty years or more in this era of increasing specialization—to find a person with knowledge of the historical lines of mathematical development.

Rota impressed me by his knowledge of some half-forgotten fields, the work of Sylvester, Cayley and others on classical invariant theory, and by the way he managed to connect the work of Italian geometers to Grassmanian geometry and modernize much of this research which dates to the last century. His main field of work was in combinatorial analysis, where again he managed to update some classical ideas and adapt them to geometry.

I suggested that Rota be invited to Los Alamos for a visit as a consultant. He has been doing this periodically ever since and proved very useful in several ways, including numerical analysis which is important in many of the large computational problems worked on the electronic computers.

Rota's personality is compatible with mine. His general education, active interest in philosophy (he is an expert on the work of Edmund Husserl and Martin Heidegger), and, above all, his knowledge of classical Latin and ancient history, have made him fill the gap left by the loss of von Neumann. Indeed we often vie in quoting from Horace, Ovid, and other authors in a good-humored display of boasting erudition. Rota is also a true bon vivant, exceedingly fond of good wines and foods, especially those from Italy. He is incredibly adept in the preparation of a great variety of pasta dishes. Italian born, he was brought to South America right after World War II, and at the age of eighteen he came to the United States. His college education took place here, but he has retained many European mannerisms in dress, tastes and habits. He is a Princeton graduate and now a professor at MIT.

CHAPTER 14

Professor Again

1967–1972

D URING the Los Alamos years I frequently took time
off to return to academic life, and around 1965 I
started visiting the University of Colorado on a
more regular basis, so it was not a discontinuous change
when in 1967 I decided to retire from Los Alamos and accept a professorship in Boulder. Nor was I going to a
strange, new place; on the contrary, I was joining several of
my good old friends who had also selected the Colorado
Rockies as a place to live, David Hawkins, Bob Richtmyer
and George Gamow. Hawkins had been a professor of philosophy in Boulder since he had left Los Alamos after the
war; Richtmyer, the post-war leader of the theoretical division before Carson Mark, had given up the Courant Institute
in New York for the cleaner air of Boulder; Gamow had
become a professor in the physics department several years
earlier. The University of Colorado was flourishing and ex-
panding, especially in the sciences, and the mathematics

department experienced an explosive growth in size and in quality. Besides, Boulder was sufficiently close to Los Alamos, an easy day's drive through spectacular scenery, so I could continue as a consultant and visit frequently. The focus of my involvement, however, shifted from Los Alamos to Boulder.

In Boulder I saw a great deal of Gamow until his death in 1968. His health had been failing for a few years, his liver had weakened under the assaults of a lifetime of carefree drinking. He was quite aware of this and said to me on some occasion: "Finally my liver is presenting me with the bill." This did not prevent him from working and writing till the very end. At his Russian funeral, when he lay in an open casket, I realized that he was only the second dead man I had ever seen in my life. Though I was not conscious of the shock this gave me, I had to hold onto the rail when we stood up for the chants so that my knees would not buckle under me.

Gamow's autobiography, *My World Line*, was published posthumously from fragments of his unfinished manuscript.

By an incredible coincidence, Gamow and Edward Condon, who had discovered simultaneously and independently the explanation of radioactivity (one in Russia, the other in this country), came to spend the last ten years of their lives within a hundred yards of each other in Boulder. They had become friends even though Condon often felt that neither he nor his collaborator Gurney had received their due share of credit for the discovery.

Condon was a marvelous person. For me he typified the best in the native American character, earthy, super-honest, solid, simple, and at the same time very perspicacious. His political views often coincided with mine. He did not like Nixon, who had hounded him on the Un-American Activities Committee to the point that he resigned from the directorship of the Bureau of Standards. He joined the physics department in Boulder after developing heart trouble; the year before his death in 1973, he had been given an artificial heart valve, which rendered his last months more active and comfortable.

In the relative greater freedom of university life, longer vacations, no fixed schedule except for some teaching, I was returning to a more academic type of science in a milieu of mathematicians and physicists. The mathematics department was acquiring excellent researchers in the foundations of mathematics, set theory, logic, and number theory. Wolfgang Schmidt, an Austrian by birth, was one of them, powerful and original in the latter. Another is a younger, brilliant Pole, Jan Mycielski, a student of Steinhaus, whom I invited to accept a professorship when I was chairman of the department. We have since collaborated in problems of game theory, combinatorics, set theory and—during the last several years—on mathematical schemata connected with the study of the nervous system. Mycielski, with Rota and a Los Alamos mathematician, William Beyer, gathered and edited the first volume of my collected works, which has been published by the MIT Press under the title *Sets, Numbers, Universes*. The Boulder mathematics department also has a number of young people strong in analysis and topology.

In 1967 the mathematician Mark Kac and I were invited by the editors of the *Encyclopaedia Britannica* to write a long article which was to be part of a series of special appendices to a new edition of the *Britannica*. Since then it has appeared separately under the title *Mathematics and Logic*. It received very favorable reviews and has been translated into French, Spanish, Russian, Czech, and Japanese. It was rather difficult for us to find the right level of presentation. Designed not so much for the broad public but rather for scientists in other fields, we tried to make it a semi-popular presentation of modern ideas and perspectives of the great concepts of mathematics.

As the reader may have noticed, much of my work seems to have been done in collaboration with others (just as this book was assembled in collaboration with Françoise). One of the reasons for this is my leaning on conversation as a stimulant to thinking; the other is my well-known impatience with detail and a certain distaste for reading what I have written. When I see one of my papers in print, I have a childish complex, a tiny nagging doubt that it might be wrong or that it may not contain anything interesting, and I discard it after a quick glance.

Mark Kac had also studied in Lwów, but since he was several years younger than I (and I had left when only twenty-six myself), I knew him then only slightly. He told me that as a young student he had been present at my doctorate ceremony and had been impressed by it. He added that these first impressions usually stay, and that he still considers me "a very senior and advanced person," even though the ratio of our ages is now very close to one. He came to America two or three years after I did. I remembered him in Poland as very slim and slight, but here he became rather rotund. I asked him, a couple of years after his arrival, how it had happened. With his characteristic good humor he replied: "Prosperity!" His ready wit and almost constant joviality make him extremely congenial.

After the war he visited Los Alamos, and we developed our scientific collaboration and friendship. After a number of years as a professor at Cornell he became a professor of mathematics at The Rockefeller Institute in New York (now The Rockefeller University.) He and the physicist George Uhlenbeck have established mathematics and physics groups at this Institute, where biological studies were the principal and almost exclusive subject before.

Mark is one of the very few mathematicians who possess a tremendous sense of what the real applications of pure mathematics are and can be; in this respect he is comparable to von Neumann. He was one of Steinhaus's best students. As an undergraduate he collaborated with him on applications of Fourier series and transform techniques to probability theory. They published several joint papers on the ideas of "independent functions." Along with Antoni Zygmund he is a great exponent and true master in this field. His work in the United States is prolific. It includes interesting results on probability methods in number theory. In a way, Kac, with his superior common sense, as a mathematician is comparable to Weisskopf and Gamow as physicists in their ability to select topics of scientific research which lie at the heart of the matter and are at the same time of conceptual simplicity. In addition—and this is perhaps related—they have the ability to present to a wider scientific audience the most recent and modern results and techniques in an understandable and often very exciting manner. Kac is a wonderful lecturer, clear, intelligent, full of sense and avoidance of trivia.

Among the mathematicians of my generation who influenced me the most in my youth were Mazur and Borsuk. Mazur I have described earlier. As for Borsuk, he represented for me the essence of geometrical intuition and truly meaningful topology. I gleaned from him, without being able to practice it myself, the workings of n-dimensional imagination. Today Borsuk is continuing his creative work in Warsaw. His recent theory of the "shape" in topology shows increasing power and applications. His general interests and mathematical outlook are very close to mine, and our old friendship was renewed after the war during his visits to the United States and my brief trip to Poland in 1973 when I saw him in his country house near Warsaw.

One could go on ad infinitum recollecting from memory, reflecting and writing down. If the reader is still with me, he may have derived from the preceding a sort of existentialist (the word is in vogue) picture of my life, these times, and the many scientists I have known. By way of a conclusion to this chapter I will add a self-portrait which I sent to Françoise before we were married. I am translating from the French which accounts in part for its awkwardness.

PROFESSOR AGAIN

"Self-portrait of Mr. S. U.

"His expression is usually ironic and quizzical. In truth he is very much affected by all that is ridiculous. Perhaps he has some talent to recognize and feel it at once, so it is not surprising that this is reflected in his facial expression.

"His conversation is very uneven, sometimes serious, sometimes gay, but never tiring or pedantic. He only tries to amuse and distract the people he likes. With the exception of the exact sciences, there is nothing which appears so certain or obvious to him that he would not allow for differing opinions: on almost any subject one can say almost anything.

"He brought to the study of mathematics a certain talent and facility which allowed him to make a name for himself at an early age. Dedicated to work and solitude until he was twenty-five, he became more worldly rather late. Nevertheless he is never rude because he is neither coarse nor hard. If he sometimes offends it is through inattention or ignorance. In speech he is neither gallant nor graceful. When he says kind things it is because he means them. Therefore the essence of his character is a frankness and truthfulness which are sometimes a little strong but never really shocking.

"Impatient and choleric to the point of violence, everything that contradicts or wounds him affects him in an uncontrollable way, but this usually disappears when he has vented his feelings.

"He is easy to influence or govern provided he is unaware that this is intended.

"Some people think that he is malicious because he makes merciless fun of pretentious bores. His temperament is naturally sensitive and renders him subject to delicate moods. This makes him at once gay and melancholy.

"Mr. U. behaves according to this general rule: he says a lot of foolish things, seldom writes them and never does any."

(271)

When Françoise read this description she felt it agreed well with what she then knew of me but was very surprised at the quality of my French, until she came to the last paragraph:

"And now I shall change from my text which I came upon by chance yesterday. The above are verbatim extracts from a letter of d'Alembert to Mademoiselle de Lespinasse written some two hundred years ago!" (D'Alembert was a famous French mathematician and encyclopedist of the eighteenth century). Françoise was very amused.

Some thirty years have elapsed since I copied this little text. I will now add as a finishing touch that I don't think I have changed much, but that there is one trait which d'Alembert did not mention that I possess—all this merely *si parva magnis comparare licet*—it is a certain impatience. I have been afflicted with this all my life. It may be increasing with advancing years. (If Einstein or Cantor came to lecture here today I would have the split reaction of a schoolboy—wanting to learn on the one hand and to skip class on the other.) While I still feel quite happy giving lectures, talks, or discussions I am becoming less and less able to sit through hours of such given by others. I am, I told some colleagues, "like an old boxer who can still dish it out but can't take it any more." This amused them no end.

CHAPTER 15

Random Reflections on Mathematics and Science

THIS chapter will be somewhat different in content from the preceding account of my "adventures" and of scientists I have known.

Here I have tried to gather, review, and sometimes amplify some of the general ideas I have touched upon so lightly throughout the book. I hope that in their randomness these reflections will give the reader an added glimpse into the manifold aspects of science and especially the relation of mathematics to other sciences. It is merely about the "gist of the gist." For greater detail, I can only refer the reader to some of my more general scientific publications.

What exactly is mathematics? Many have tried but nobody has really succeeded in defining mathematics; it is always something else. Roughly speaking, people know that it deals with numbers and figures, with patterns, relations, operations, and that its formal procedures involving axioms, proofs, lemmas, theorems have not changed since the time of Archimedes. They also know that it purports to form the foundations of all rational thought.

Some could say it is the external world which has molded our thinking—that is, the operation of the human brain—into what is now called logic. Others—philosophers and scientists alike—say that our logical thought (thinking process?) is a creation of the internal workings of the mind as they developed through evolution "independently" of the action of the outside world. Obviously, mathematics is some of both. It seems to be a language both for the description of the external world, and possibly even more so for the analysis of ourselves. In its evolution from a more primitive nervous system, the brain, as an organ with ten or more billion neurons and many more connections between them must have changed and grown as a result of many accidents.

The very existence of mathematics is due to the fact that there exist statements or theorems, which are very simple to state but whose proofs demand pages of explanations. Nobody knows why this should be so. The simplicity of many of these statements has both aesthetic value and philosophical interest.

The aesthetic side of mathematics has been of overwhelming importance throughout its growth. It is not so much whether a theorem is useful that matters, but how elegant it is. Few non-mathematicians, even among other scientists, can fully appreciate the aesthetic value of mathematics, but for the practitioners it is undeniable. One can, however, look conversely at what might be called the homely side of mathematics. This homeliness has to do with having to be punctilious, of having to make sure of every step. In mathematics one cannot stop at drawing with a big, wide brush; all the details have to be filled in at some time.

RANDOM REFLECTIONS

"Mathematics is a language in which one cannot express unprecise or nebulous thoughts," said Poincaré, I believe in a speech on world science which he gave at the St. Louis fair many years ago. And he gave as an example of the influence of language on thought a description of how differently he felt using English instead of French.

I tend to agree with him. It is a truism to say that there is a clarity to French which is not there in other tongues, and I suppose this makes a difference in the mathematical and scientific literature. Thoughts are steered in different ways. In French generalizations come to my mind and stimulate me toward conciseness and simplification. In English one sees the practical sense; German tends to make one go for a depth which is not always there.

In Polish and Russian, the language lends itself to a sort of brewing, a development of thought like tea growing stronger and stronger. Slavic languages tend to be pensive, soulful, expansive, more psychological than philosophical, but not nebulous or carried by words as much as German, where words and syllables concatenate. They concatenate thoughts which sometimes do not go very well together. Latin is something else again. It is orderly; clarity is always there; words are separated; they do not glue together as in German; it is like well-cooked rice compared to overcooked.

Generally speaking, my own impressions of languages are the following: When I speak German everything I say seems overstated, in English on the contrary it feels like an understatement. Only in French does it seem just right, and in Polish, too, since it is my native language and feels so natural.

Some French mathematicians used to manage to write in a more fluent style without stating too many definite theorems. This was more agreeable than the present style of the research papers or books which have so much symbolism and formulae on every page. I am turned off when I see only formulas and symbols, and little text. It is too laborious for me to look at such pages not knowing what to concentrate on. I wonder how many other mathematicians really read them in detail and enjoy them.

There do exist, though, important, laborious and inelegant theorems. For example some of the work connected with partial differential equations tends to be less "beautiful" in form and style, but it may have "depth," and may be pregnant with consequences for interpretations in physics.

How does one arrive at a value judgment nowadays?

Mathematicians, whose job in a sense is to analyze the motivation and origin of their work, fool themselves and may be remiss when they think their main business is to prove theorems without at least indicating why they may be important. If left entirely to aesthetic criteria, doesn't it compound the mystery?

I believe that in the decades to come there will be more understanding, even on a formal level, of the degree of beauty, though by that time the criteria may have shifted and there will be a super beauty in unanalyzable higher levels. So far when anyone has tried to analyze the aesthetic criteria of mathematics too precisely, whatever was proposed has seemed too narrow. It has to appeal to connections with other theories of the external world or to the history of the development of the human brain, or else it is purely aesthetic and very subjective in the sense that music is. I believe that even the quality of music will be analyzable—to an extent only, of course—at least by formal criteria, by mathematizing the idea of analogy.

Some of the old problems, unsolved for many years, are being settled. Some are solved with a bang, and others with a whimper, so to speak. This applies to problems seemingly equally important and a priori interesting, but some, even famous classical ones, are solved in such a specific way that there is nothing more to be asked or said. Some others, less famous, immediately upon solution become sources of curiosity and activity. They seem to open new vistas.

As for publications, mathematicians nowadays are almost

forced to conceal the way they obtain their results. Evariste Galois, the young French genius who died at the age of twenty-one, in his last letter written before his fatal duel, stressed how the real process of discovery is different from what finally appears in print as the process of proof. It is important to repeat this again and again.

On the whole and in the large lines there does seem to exist a consensus among working mathematicians about the value of individual achievements and the value of new theories. There must therefore be something objective if not yet defined about the feeling of beauty which mathematics offers, dependent sometimes also on how useful it turns out to be in other branches of itself or of other sciences. Why mathematics is really so useful in the description of the physical world, for me at least remains philosophically a mystery. Eugene Wigner once wrote a fascinating article on this "implausible" usefulness of mathematics and titled it "The Unreasonable Effectiveness of Mathematics."

It is, of course, one very concise way of formalizing all rational thought.

It also has manifestly, in elementary, secondary and advanced schools the value of training our brain, since practice, just as in any other game, sharpens the organ. I cannot say whether a mathematician's brain is today sharper than it was in the time of the Greeks; nevertheless on the longer scale of evolution it must be so. I do believe mathematics may have a great genetic role, it may be one of the few means of perfecting the human brain. If true, nothing could then be more important for humanity, whether to arrive at some new destiny as a group or for individuals. Mathematics may be a way of developing physically, that is anatomically, new connections in the brain. It has a sharpening value even though the enormous proliferation of material shows a tendency to beat things to death.

Yet every formalism, every algorithm, has a certain magic in it. The Jewish Talmud, or even the Kabbalah, contains material which does not appear particularly enlightening intellectually, being a vast collection of grammatical or culinary recipes, some perhaps poetic, others mystical, all rather arbitrary. Over centuries thousands of minds have pored over, memorized, dissected, and classified these works. In so doing people may have sharpened their memories and deductive practice. As one sharpens a knife on a whetstone, the brain can be sharpened on dull objects of thought. Every form of assiduous thinking has its value.

There exist in mathematics propositions, such as the one called "Fermat's Great Theorem," which, standing by themselves, seem special and unrelated to the main body of number theory. They are very simple to state but have defied all the efforts of the greatest minds to prove them. Such statements have stimulated young minds (my own included) to more general wonder and curiosity. In the case of Fermat's problem, special or irrelevant as it is by itself, it has stimulated through the last three centuries of mathematics, the creation of new living objects of mathematical thought, in particular the so-called theory of ideals in algebraic structures. The history of mathematics knows a number of such creations.

The invention of imaginary and complex numbers (which are pairs of real numbers with a special rule for their addition and multiplication) beyond the immediate purpose and the use to which they were put, opened new possibilities and led to the discovery of miraculous properties of the complex variables. These analytic functions (the examples of which are, to mention the simplest, $z = \sqrt{w}$, $z = e^w$, $z = \log w$), possess unexpected, simple and a priori unforeseen properties deriving from the few general rules which govern them. They have convenient algorithms and rather deep connections with the properties of geometrical objects and also with the mysteries concerning the seemingly so familiar natural numbers, the ordinary integers. It is as if some invisible different universe governing our thought became dimly perceptible through it, a universe with some laws, and yes, facts, of which we become only vaguely aware.

The fact that some seemingly very special functions, like the Riemann Zeta function, have such deep connections with the behavior of integers, of prime numbers, is hard to explain a priori and in depth. This is really not well understood to this day. Somehow these entities, these special analytic functions defined by infinite series, have been generalized more recently to spaces other than the plane of all complex numbers, such as to algebraic surfaces. These entities show connections between seemingly diverse notions. They also seem to show the existence (to make a metaphor stimulated by the subject itself) of another surface of reality, another Riemann surface of thought (and connections of thought) of which we are not consciously aware.

Some of the properties of the analytic functions of the complex numbers turn out to be not merely convenient, but very fundamentally tied to physical properties of matter, in the theory of hydrodynamics, in the description of the motions of incompressible fluids such as water, in electrodynamics, and in the foundations of quantum theory itself.

The creation of a general idea of a space, abstracted to be sure, but not really completely dictated or uniquely indicated by the physical space of our senses, the generalization to the n-dimensional space where n is greater than three and even to infinitely many dimensions, and so useful at least as a language for the foundations of physics itself, are these marvels of the power of the human brain? Or is it the nature of the physical reality which reveals it to us? The very invention, or is it "discovery," that there are different degrees or different kinds of infinity has had not only a philosophical, but beyond that, a striking psychological influence on receptive minds.

Speaking of the fascination of surprises, the mysterious attraction of mathematics and, of course, of other sciences—

physics especially—it may be remarked how often it happens that in the game of chess one may observe weak players or even rank beginners getting into deep and fascinating positions. I have often watched amateurs or nontalented beginners, looked at their game after some fifteen moves, observed that their position arrived at perhaps by chance, certainly not by design, was full of marvelous possibilities for both sides. And I wonder how it is that the game itself produces these positions of great appeal and art without these simple fellows being even aware of it. I do not know whether an analogous experience is possible in the game of Go. I cannot myself judge, not knowing much about the intricacies of that beautiful game, but I wonder whether a master looking at a position can tell whether it was arrived at by chance or by a logically developed correct and thoughtful play.

In science, and in mathematics in particular, there seems to be a similar magical interest in certain algorithms. They appear to have a power to produce by themselves, as it were, solutions to problems or vistas of new perspectives. What seemed to be at first mere tools designed for special purposes can bring about some unforeseen and unexpected new uses.

By the way, a little philosophical conundrum occurred to me which I do not know how to resolve: Consider a game like a solitaire, or a game between two persons. Assume that the players may cheat once or twice during the course of the game. For instance, in a Canfield solitaire, if one changes the position of one or two cards once and once only, the game is not destroyed. It would still be a precisely, completely, mathematically meaningful, albeit different game. It would become simply a bit richer, more general. But if one takes a mathematical system, a system of axioms and allows the addition of one or two false statements, the result is immediate nonsense because once one has a false statement, one can deduce anything one wants to. Where does the difference lie? Perhaps it lies in the fact that only a certain class of motions is allowed in the game, whereas in mathematics once an incorrect statement is introduced one may immediately get the statement: zero equals one. There must then be a way to generalize the game of mathematics so that one could make a few mistakes and instead of getting complete nonsense, obtain merely a wider system.

Hawkins and I have speculated on the following related problem: a variation on the game of Twenty Questions. Someone thinks of a number between one and one million (which is just less than 2^{20}). Another person is allowed to ask up to twenty questions, to each of which the first person is supposed to answer only yes or no. Obviously the number can be guessed by asking first: Is the number in the first half million? then again reduce the reservoir of numbers in the next question by one-half, and so on. Finally the number is obtained in less than $\log_2 (1,000,000)$. Now suppose one were allowed to lie once or twice, then how many questions would one need to get the right answer? One clearly needs more than n questions for guessing one of 2^n objects because one does not know when the lie was told. This problem is not solved in general.

In my book on unsolved problems I claim that many mathematical theorems can be payzised (a Greek word which means to play). That is, that they can be formulated in game-theoretical language. For example, a rather general schema for playing a game can be set up as follows:

Suppose N is a given integer and two players are to build two permutations of N letters $(n_1, n_2, \ldots n_N)$. They are constructed by the two players in turn, as follows. First permutation, the first player takes n_1, the second n_2, the first takes n_3 and so on. Finally the first permutation is accomplished. Then they play for the second permutation, and if the two permutations generate the group of all permutations, the first player wins, if not the second wins. Who has a winning strategy in this game? This is merely a small example of how, in any domain of mathematics—in this case in finite group theory—one can invent gamelike schemata which lead to purely mathematical problems and theorems.

One can also ask questions of a different type: If this is done at random what are the chances? This is a problem that combines measure theory, probability and combinatorics. One may proceed this way in many domains of mathematics.

Set theory revolutionized mathematics toward the end of the nineteenth century. What started this was that Georg Cantor proved (i.e., discovered) that the continuum is not countable. He did have predecessors in these speculations on the logic of infinity, Weierstrass and Bolzano, but the first precise study of degrees of infinity was certainly his. This arose from his discussion of trigonometric series, and very quickly transformed the shape and flavor of mathematics. Its spirit has increasingly pervaded mathematics; recently it has had a new and technically unexpected renewed development, not only in its most abstract form, but also in its immediate applications. The formulations of topology, of algebraical ideas in their most general form received impetus and direction from the activities of the Polish school, much of it from Lwów where the interests centered around what can be called roughly functional analysis in a geometrical and algebraical spirit.

To give an oversimplified description of the origin of much of these activities: After Cantor and the mathematicians of the French school, Borel, Lebesgue and others, this kind of investigation found a home in Poland. In her book *Illustrious Immigrants*, Laura Fermi expresses a surprised admiration at the large percentage of Polish mathematicians in the United States who contributed so much significant work to the flourishing of this field. Many came here to settle and continue such work. Simultaneously the studies of analysis of Hilbert and other German mathematicians brought about a simple general mathematical construction of infinitely dimensional functional spaces, also later further developed by the Polish school. Independently and at the same time, the work of Moore, Veblen, and others in America brought about a meeting of the geometrical and algebraical points of view, and a unification—only to some extent to be sure—of mathematical activities.

It appears that in spite of increasing diversity and even overspecialization, the choice of mathematical topics of research thus follows prevalent currents, threads, and trends which come together from independent sources.

Some few individuals with a few new definitions are apparently able to start a whole avalanche of work in special fields. This is partly due to fashion and self-perpetuation by the sheer force of the teachers' influence. When I first came to this country I was amazed at what seemed to me an exaggerated concentration on topology. Now I feel there is perhaps too much work in the domain of algebraic geometry.

A second epochal landmark was Gödel's work, recently made more specific by Paul Cohen's results. Gödel, the mathematical logician at the Institute for Advanced Studies in Princeton, found that any finite system of axioms or even countably infinite systems of axioms in mathematics, allows one to formulate meaningful statements within the system which are undecidable—that is to say, within the system one will not be able to prove or disprove the truth of these statements. Cohen opened the door to a whole class of new axioms of infinities. There is now a plethora of results showing that our intuition of infinity is not complete. They open up mysterious areas in our intuitions to different concepts of infinity. This will, in turn, contribute indirectly to a change in the philosophy of foundations of mathematics, indicating that mathematics is not a finished object as was believed, based on fixed, uniquely given laws, but that it is genetically evolving. This point of view has not yet been accepted consciously, but it points a way to a different outlook. Mathematics really thrives on the infinite, and who can tell what will happen to our attitudes toward this notion during the next fifty years? Certainly, there will be something—if not axioms in the present sense of the word, at least new rules or agreements among mathematicians about the assumption of new postulates or rather let us call them formalized desiderata, expressing an absolute freedom of thought, freedom of construction, given an undecidable proposition, in preference to true or false assumption. Indeed some statements may be undecidably undecidable. This should have great philosophical interest.

The interest in the foundations of mathematics is to some extent also philosophical, though eventually it does pervade everything, like set theory. But the word "foundations" is a misnomer; for the time being, it is just one more mathematical specialty, fundamental to be sure.

The great dichotomy in the origin and in the inspiration of mathematical thought—stimulated by the influence of external reality, the physical universe, on one hand, and by the developing processes of physiology, almost perhaps of the human brain, on the other—has in a small and special way a homomorphic image in the present and near-future use of electronic computers.

Even the most idealistic point of view about mathematics as a pure creation of the human mind must be reconciled with the fact that the choice of definitions and axioms of geometry—in fact of most mathematical concepts—is the result of impressions obtained through our senses from external stimuli and inherently from observations and experiments in the "external world." The theory of probability, for example, came about as a development of a few questions concerning games of chance. Now, computing machines constructed to solve special problems of mathematics promise to enlarge very greatly the scope of the Gedanken Experimente, the idealization of experience, and our more abstract schemata of thought.

It appears that experimentation on models of games played by self-organizing living matter through chemical reactions in living organisms will lead to novel abstract mathematical schemata. The new study of the mathematics of growing patterns, and the possibility of studying experimentally on computing machines the course of competitions or contests between geometrical configurations imitating the fight for survival, these might give rise to new mathematical setups. One could again give names like "payzonomy" to the combinatorics of contesting reactions and "auxology" to a yet-to-be-developed theory of growth and organization, this latter ultimately including the growing tree of mathematics itself.

So far only the very simplest and crude mathematical schemata have been proposed to mirror the mathematical properties of geometrical growth. (An account of my own simple-minded models can be found in a recent book edited by Arthur Burke, *A Theory of Cellular Automata*, published by the University of Illinois Press.)

An especially ingenious set of rules was devised by the English mathematician John Conway, a number theorist. The Conway Game of Life is an example of a game or pastime which, perhaps much like the early problems involving dice and cards, has led ultimately to the present edifice of probability theory, and may lead to a vast new theory describing the "processes" which Alfred North Whitehead studied in his philosophy.

The use of computers seems thus not merely convenient, but absolutely essential for such experiments which involve following the games or contests through a very great number of moves or stages. I believe that the experience gained as a result of following the behavior of such processes will have a fundamental influence on whatever may ultimately generalize or perhaps even replace in mathematics our present exclusive immersion in the formal axiomatic method.

The already-mentioned recent results of Paul Cohen and others—Peter Novikoff, Hao Wang, Yuri Matiasević—on the independence from the traditional system of axioms of some of the most fundamental mathematical statements, indicate a new role for pragmatic approaches. Work with automata will help indicate whether a problem can be solved by existing means.

To illustrate what we have in mind let us consider for example a "little" special problem in three dimensions: given a closed curve in space and a solid body of given shape, the problem is to push the body through the curve. There are no clear mathematical criteria to decide whether it can be done or not. One has to rotate, wiggle, squeeze, and "try," to see whether it can be done. In a higher number of dimensions, like five, one can have an analogous problem. The idea is to set it up on a computer and try various possible motions. Perhaps, after very many tries, one would acquire a feeling for the freedom of maneuvering in this high dimensional space and a new type of an almost tactile intuition. Of course, this is a special, small and unimportant example, but I feel that one could develop new imaginations by suitable experimentation with these new tools, electronic computers especially, in setting up and observing the various growth processes and evolutionary developments.

It seems to me the impact and role of the electronic computer will significantly affect pure mathematics also, just as it has already done so in the mathematical sciences, principally in physics, astronomy and chemistry.

These conjectured excursions into aspects of the future of mathematics take us far from von Neumann and his contemporaries, and their role in the evolution of science a quarter-century ago. The rate of growth in the organized activities of the human mind, accelerated no doubt by the advent of computers, seems to increase in a way which forebodes qualitative changes in our way of thinking and living. As Niels Bohr said in one of his amusing remarks: "It is very hard to predict, especially the future." But I think mathematics will greatly change its aspect. Something drastic may evolve, an entirely different point of view on the axiomatic method itself. Instead of detailed work on special theorems which now number in the millions, instead of thinking in terms of rules operating with symbols given once and for all, it may be that mathematics will consist more and more of problems, or desiderata, or programs for work of a general nature. No longer will there be additional multitudes of special spaces, definitions of special manifolds, of special mappings of this and that—though a few will survive: "apparent rari nantes in gurgite vasto," no new collections of individual theorems, but instead general sketches or outlines of larger theories, of vaster enterprises, and the actual working out of proofs of theorems will be left to students or even to machines. It may become comparable to impressionistic painting in contrast to the painful, detailed drawing of earlier days. It could be a more living and changing scene, not only in the choice of definitions but in the very rules of the game, this great game whose rules until now have not changed since antiquity.

If the rules have not changed, very great changes have already taken place in the scope of mathematics in the space of my own lifetime. In the nineteenth century the applications of mathematics were all-inclusive in physics, astronomy, chemistry, in mechanics, engineering, and all the other facets of technology. More recently, mathematics serves to formulate the foundations of other sciences as well, so-called mathematical physics is really the theory of all physics, reaching into its most abstract parts like quantum theory, the very strange four-dimensional continuum of space-time. These belong specifically to the twentieth century. In the short span of sixty to one hundred years the proliferation of the use of mathematical ideas has been unbelievably varied. It was accompanied by, one could say, an explosive creation of new mathematical objects, large and small, and a tendency to "beat things to death" with proliferation and hairsplitting studies of minute details that is almost Talmudic.

At a talk which I gave at a celebration of the twenty-fifth anniversary of the construction of von Neumann's computer in Princeton a few years ago, I suddenly started estimating silently in my mind how many theorems are published yearly in mathematical journals. (A theorem being defined as a statement which is just labeled "theorem," and is published in a recognized mathematical journal.) I made a quick mental calculation, amazing myself that I could do this while talking about something entirely different and came to a number like one hundred thousand theorems per year. Quickly changing my topic I mentioned this and the audience gasped. It may interest the reader that the next day two of the younger mathematicians in the audience came to tell me that impressed by this enormous figure they undertook a more systematic and detailed search in the Institute library. By multiplying the number of journals by the number of yearly issues, by the number of papers per issue and the average number of theorems per paper, their estimate came to nearer two hundred thousand theorems a year. Such an enormous number should certainly give food for thought. If one believes that mathematics is more than games and puzzles, here is something to worry about. Clearly the danger is that mathematics itself will suffer the fate of splitting into different separate sciences, into many independent disciplines tenuously connected. My own hope is that this will not happen, for if the number of theorems is larger than one can possibly survey, who can be trusted to judge what is "important"? The problem becomes one of record keeping, of storage and retrieval of the results obtained. This problem now becomes paramount; one cannot have survival of the fittest if there is no interaction.

It is actually impossible to keep abreast of even the more outstanding and exciting results. How can one reconcile this with the view that mathematics will survive as a single science? Just as one cannot know all the beautiful women or all the beautiful works of art and one finally marries one beautiful person, one can say that in mathematics one becomes married to one's own little field. Because of this, the judgment of value in mathematical research is becoming more and more difficult, and most of us are becoming mainly technicians. The variety of objects worked on by young scientists is growing exponentially. Perhaps one should not call it a pollution of thought; it is possibly a mirror of the prodigality of nature which produces a million species of different insects. Somehow one feels, though, that it goes against the grain of one's ideals of science, which aims to understand, abbreviate, summarize, and, in particular, to develop, a notation system for the phenomena of the mind and of nature.

It is the unexpected in the development of science, the way really new ideas and concepts strike a young mind, that mold it irreversibly. Later, for the mature or older mind, the unexpected causes a wonder which induces new stimulation, even when one has become less impressionable or even jaded. To quote Einstein, "The most beautiful thing we can experience is the mysterious. It is the source of all true art and science."

Mathematics creates new objects of thought—one could call it a meta-reality—by engendering ideas which begin to live their own life in an independent development. Once born these cannot any more be controlled by a single person, only by a collection of brains which are the perpetuating set of mathematicians.

Talent or genius in mathematics is hard to quantify. I tend to feel that there is an almost continuous passage from mediocrity to the highest levels of people like Gauss, Poincaré and Hilbert. So much depends not on the brain alone. There are definitely what I have called, for want of a better word, "hormonal factors" or traits of character: stubbornness, physical ability, willingness to work, what some call "passion." These depend a great deal on habits mostly acquired in childhood or early youth when accidents of early impressions play a great role. Undoubtedly, much of the quality called imagination or intuition comes from the physiological structural properties of the brain, which in turn may be partly developed through experiences leading to certain habits of thought and of the direction of the train of thought.

The willingness to plunge into the unknown and the unfamiliar varies with different individuals. There are distinctly different types of mathematicians—those who prefer to attack existing problems or to build on what is already there, and those who like to imagine new schemata and new possibilities. The first perhaps constitute a majority, maybe more than eighty percent. When a young man wants to establish a reputation he will mostly attack an unsolved problem that has already been worked on. In this way, if he is lucky or strong enough, it will be comparable to an athlete beating a record, jumping higher than anyone before. Although what is often of greater value is the conception of a new idea, a young person is often unwilling to try this, not knowing whether the new thought will be appreciated even when to him it is important and beautiful.

I am of the type that likes to start new things rather than improve or elaborate. The simpler and "lower" I can start the better I like it. I do not remember using complicated theorems to prove more complicated ones. (Of course, this is all relative, "there's nothing new under the sun"— everything can be traced back to Archimedes or even earlier.)

I also believe that changing fields of work during one's life is rejuvenating. If one stays too much with the same subfields or the same narrow class of problems a sort of self-poisoning prevents acquisition of new points of view and one may become stale. Unfortunately, this is not uncommon in mathematical creativity.

With all its grandiose vistas, appreciation of beauty, and vision of new realities, mathematics has an addictive property which is less obvious or healthy. It is perhaps akin to the action of some chemical drugs. The smallest puzzle, immediately recognizable as trivial or repetitive can exert such an addictive influence. One can get drawn in by starting to solve such puzzles. I remember when the *Mathematical Monthly* occasionally published problems sent in by a French geometer concerning banal arrangements of circles, lines and triangles on the plane. "Belanglos," as the Germans say, but nevertheless these figures could draw you in once you started to think about how to solve them, even when realizing all the time that a solution could hardly lead to more exciting or more general topics. This is much in contrast to what I said about the history of Fermat's theorem, which led to the creation of vast new algebraical concepts. The difference lies perhaps in that little problems can be solved with a moderate effort whereas Fermat's is still unsolved and a continuing challenge. Nevertheless both types of mathematical curiosities have a strongly addictive quality for the would-be mathematician which exists on all levels from trivia to the most inspiring aspects.

In the past there always were a few mathematicians who either explicitly or by implication gave specific ideas and choice of direction to the work of others—men like Poincaré, Hilbert and Weyl. This is now becoming increasingly difficult if not impossible. There is probably not one mathematician now living who can even understand all of what is written today.

A volume written more than thirty years ago by Eric Temple Bell, *The Development of Mathematics*, contains an excellent abbreviated account of the history of mathematics. (Perhaps I like it, because, to use G.-C. Rota's language, my work is mentioned there even though the book was written when I was only twenty-eight years old and it is a rather small volume. There is more satisfaction in being mentioned in a short history than in one which has ten thousand pages!) But when Weyl was asked by a publisher to write a history of mathematics in the twentieth century he turned it down because he felt that no one person could do it.

Von Neumann, who could have aspired to such a role, admitted to me some thirty-five years ago that he knew less than a third of the corpus of mathematics. At his suggestion once I concocted for him a doctoral-style examination in various fields trying to select questions which he would not be able to answer. I did find some, one each in differential geometry, in number theory, in algebra, which he could not answer satisfactorily. (This by the way may also tend to show that doctoral exams have little permanent meaning.)

As for myself, I cannot claim that I know much of the technical material of mathematics. What I may have is a feeling for the gist, or maybe only the gist of the gist, in a number of its fields. It is possible to have this knack for guessing or feeling what is likely to be new or already known, or else not known, in some branch of mathematics where one does not know the details. I think I have this ability to a degree and can often tell whether a theorem is known, i.e. already proved, or is a new conjecture. This is a sort of feeling that comes from the way the quantifiers are arranged, from the tone or the music of the statement, so to speak.

Speaking of this analogy: I can remember tunes and am able to whistle various melodies rather correctly. But when I try to invent or compose some new "catchy" tune, I find rather impotently that what I do is a trivial combination of what I have heard. This in complete contrast to mathematics where I believe with a mere "touch" I can always propose something new.

Collaboration in mathematics is a very interesting and new phenomenon which developed during the last several decades.

It is natural in experimental physics that investigators work together on the different phases of instrumentation. By now every experiment is really a class of technical projects, especially on the great machines which require hundreds of engineers and specialists for their construction and operation. In theoretical physics this is perhaps not as evident, but it exists, and strangely enough in mathematics also. We have seen that the creative effort in mathematics requires intense concentration and constant thinking in depth for hours on end, and that it is often shared by two individuals who just look at each other and occasionally make a few remarks when they collaborate. It is now definitely so that even in the most abstruse mathematical questions two or more persons work together on trying to find a proof. Many papers have now two, sometimes three or more authors. The exchange of conjectures, suggesting tentative approaches, helps to build up partial results along the way. It is easier to talk than to write down every thought. There is here an analogy to analyzing a game of chess.

It may be that in the future large groups of mathematicians working together will produce important, beautiful, and simple results. Some have already been produced this way in recent years. For example, the solution of one of Hilbert's problems about the existence of algorithms to solve diaphantine equations was really obtained (not in parallel to be sure but in sequence) by several scientists in this country, and at the end by a young Russian, Yuri Matiasević, who took the last step. Several mathematicians working independently in the United States and in Poland but aware of each other's results, solved an old problem of Banach's about the homeomorphism of his spaces. They were able to climb on each other's shoulders, so to speak.

It was after the publicity surrounding the construction of the atomic bomb in Los Alamos that the expression "critical mass" became current as a metaphoric description of the required minimal size of a group of scientists working together in order to obtain successful results. If large enough, the group produces results explosively. When the critical mass is reached, due to mutual stimulation the multiplication of results, like that of neutrons, becomes exponentially larger and more rapid. Before such a mass is attained, progress is gradual, slow and linear.

Other variations in the working habits of scientists have been slower. The mode of life in the ivory tower world of science now includes more scientific meetings, more involvement in governmental work.

A simple but important thing like letter-writing has also undergone a noticeable change. It used to be an art, not only in the world of literature. Mathematicians were voluminous letter writers. They wrote in longhand and communicated at length intimate and personal details as well as mathematical thoughts. Today the availability of secretarial help renders such personal exchanges more awkward, and as it is difficult to dictate technical material scientists in general and mathematicians in particular exchange fewer letters. In my file of letters from all the scientists I have known, a collection extending over more than forty years, one can see the gradual, and after the war accelerated, change from long, personal, handwritten letters to more official, dry, typewritten notes. In my correspondence of recent years, only two persons have continued to write in longhand: George Gamow and Paul Erdös.

Chen Ning Yang, the Nobel prize physicist, tells a story which illustrates an aspect of the intellectual relation between mathematicians and physicists at present:

One evening a group of men came to a town. They needed to have their laundry done so they walked around the city streets trying to find a laundry. They found a place with the sign in the window, "Laundry Taken in Here." One of them asked: "May we leave our laundry with you?" The proprietor said: "No. We don't do laundry here." "How come?" the visitor asked. "There is such a sign in your window." "Here we make signs," was the reply. This is somewhat the case with mathematicians. They are the makers of signs which they hope will fit all contingencies. Yet physicists have created a lot of mathematics.

In some of the more concrete parts of mathematics—for example probability theory—physicists like Einstein and Smoluchowski have opened certain new areas even before mathematicians. The ideas of information theory, of entropy of information and its role in general continuum originated with physicists like Leo Szilard and an engineer, Claude Shannon, and not with "pure" mathematicians who could and ought to have done so long before. Entropy, a property of a distribution, was a notion originating in thermodynamics and was applied to physical objects. But Szilard (in very general terms) and Shannon defined this notion for general mathematical systems. True, Norbert Wiener had some part in the origin of it and wonderful mathematicians like Andrei Kolmogoroff later developed, generalized, and applied it to purely mathematical problems.

In the past some mathematicians, Poincaré for example, knew a lot of physics. Hilbert did not seem to have too much true physical instinct, but he wrote very important papers about the techniques and the logic of physics. Von Neumann knew a good deal of physics too, but I would say that he did not have the physicist's natural feeling for and recourse to experiment. He was interested in the foundations of quantum mechanics as long as they could be mathematized. The axiomatic approach to physical theories is to physics what grammar is to literature. Such mathematical clarity need not be conceptually crucial for physics.

On the other hand, much of the apparatus of theoretical physics and occasionally some precursor ideas came from pure mathematics. The general non-Euclidean geometries prophetically envisaged by Riemann as having future importance for physics, came before general relativity, and the definition and study of operators in Hilbert space came before quantum mechanics. The word spectrum, for example, was used by mathematicians long before anybody would have dreamed of using the spectrum representation of Hilbert space operators to explain the actual spectrum of light emitted by atoms.

I have often wondered why mathematicians have not generalized the special theory of relativity into different types of "special relativities," so to speak (not into the presently known general theory of relativity). I am sure there are other "relativities" possible in general spaces, yet hardly anything has been attempted by mathematicians. Endless papers exist on metric spaces generalizing the ordinary geometry without the dimension of time in it. Put in time and space together, and mathematicians stay out! Topologists stay with spatial spaces; they have not considered ideas which would generalize the four-dimensional time-space. This is very curious to me, epistemologically and psychologically. (I can think of one paper by van Danzig, which speculates philosophically around the notion of time topology; he says it might be a solenoidal variable. I like this, but clearly one should do much more imaginative work with time-like spaces.)

As is well known, the theory of special relativity postulates and is built entirely on the fact that light always has the same velocity regardless of the motion of the source or the observer. From this postulate alone everything follows, including the famous formula $E = mc^2$. Mathematically speaking, the invariance of the cones of light lead to the Lorentz group of transformations. Now a mathematician could, just for mathematical fun, postulate that the frequency, for example, remains the same, or that some other class of simple physical relation is invariant. By following logically one could see what the consequences would be in such a picture of a not "real" universe.

Mathematics is now so completely different from what it was in the nineteenth century, even if ninety-nine percent of mathematicians have no feeling for physics. There are so many ideas in physics begging for mathematical inspiration—new formulations, new mathematical ideas. I do not mean the use of mathematics in physics, but the other way round: physics as a stimulant for new mathematical concepts.

Contrary to mathematics, in physics one can, in principle, keep more abreast of what is going on in research. Every physicist can know the gist of most of physics. There are very few fundamental problems now such as the problem of the nature of elementary particles or what is the nature of the physical space and time.

In present-day research in theoretical physics, even though many of the young people are very clever, ingenious, and technically superb, their fundamental ideas tend to be orthodox, and on the whole only small variations on what has been done are produced, elaboration of details, and continuation along lines that already have been started.

Perhaps this has always been so and really new ideas are exceptional.

Sometimes, half in jest, to needle contemporary young physicist friends who spend all their time examining a few very strange particles, I tell them that it is not necessarily the best way to get new inspiration about the foundations of physics and the scheme of things in space and time.

Of course, it is not a precise problem or recognized as such, but what to my mind is a first question in physics is whether there exists a true infinity of structures going down into smaller and smaller dimensions. If so it would be worthwhile for mathematicians to speculate on whether space and time change, even in their topology, in smaller and smaller regions. We had in physics an atomistic or field-structure base. If the ultimate reality consists of a field, then its points are true mathematical points and indistinguishable. There is a possibility that in reality we have a strange structure of infinitely many stages, each stage different in nature. This is a fascinating picture which becomes more physical and not merely a philosophical conundrum. Recent experiments show definitely the increasing complication of structures. In a single nucleon we may have partons, as Feynman calls them. These partons may be the hypothetical quarks or other structures. The recent theoretical attempts no longer explain the experimental models by simple quarks, but one has to involve colored quarks of different types. Perhaps one has reached a point where it might be preferable to consider the succession of structures as going on ad infinitum.

Theoretical physics is possible because there are many identical or nearly identical copies of objects and situations. If one takes the universe by definition as only one (even though it is true that galaxies resemble each other) and the world as a whole being one, the questions asked about the cosmos as a whole have a different character. The stability with respect to adding a few more elements to an already

(297)

large number is no longer guaranteed. We have no way to observe or experiment with a number of universes. Therefore problems of cosmology and cosmogony have a different character from those of even the most fundamental physics.

Science would not be possible, physics would not be conceivable, if there was not this similarity or identity of vast numbers of points or subsets or groups of points in this universe. All individual protons seem to resemble each other, all electrons seem to resemble each other, the attraction between any two celestial bodies seems to be similar, depending only on distance and mass. So the role of physics appears, inter alia, to divide the existing groupings into entities of which there are very many examples that are isomorphic or almost isomorphic to each other. The hope for physics lies in the fact that one can almost repeat situations, or if not exactly repeat, the addition of one or more small changes makes relatively little difference. Whether there are twenty or twenty-two bodies does not make their behavior change radically. A belief in some fundamental stability! Somehow the hope is to describe physics in terms of simpler entities and identity of parts by some kind of union or counting. For example, physicists believed, at least until recently, that if one had many points, the behavior of their mass could be explained by two-body interactions—this means adding up the potentials between any two bodies. Otherwise, if every time through adding a few bodies one changed the behavior of the whole system, there would be no science of physics. This point is not sufficiently brought out in physics textbooks.

One can relate the notion of entropy to the notion of complexity, if one defines the distance between two algebraical structures and the total work necessary to prove a statement or a theorem as energy. Results exist stating that in given systems in order to prove such and such formula one needs so many steps. The minimum sufficient number of steps can be defined as an analogue of work or energy. This is worth thinking about. To make a sensible theory of it requires erudition, imagination and common sense. There is no axiomatic system even for the presently established body of physics.

Just as in pure mathematics, in theoretical physics we can see a dichotomy between the great new "unexpected" ideas and the great syntheses of established theories. Such syntheses are in a sense complementary or opposed to the new concepts. They summarize previous theories in a non-obvious way. Let me illustrate this distinction: the special theory of relativity is a priori a very strange and mysterious concept. It involves an almost irrational insight and an a priori implausible axiom based on the experimental fact that light velocity seems to be the same for a moving observer from a fixed emission point or vice versa. When the emission point moves away or toward the observer the velocity of light relative to the observer is the same no matter what the relative velocity is. From this alone a great theoretical edifice was built, a physical theory of space and time with so many surprising—and as we now know—technologically shattering consequences.

Quantum theory involves similarly, in a way, an a priori non-intuitive or unexpected set of concepts.

Maxwell's theory of electromagnetism would be an example of a great synthesis. It came after a great number of experimental facts were developed which were perhaps not so strange to their first discoverers. The theory that explains these observational facts in one set of mathematical equations constitutes one of the most impressive achievements of human thought. Epistemologically this theory is of a different nature, or so it seems to me, from relativity and quantum theory which were, one might say, more unexpected.

In astronomy, the recent observational discoveries show the continuing strangeness of the cosmos in the variety of different types of stars, of conglomerations of stars, of clusters, galaxies, and strange new objects. These include neutron stars, black holes, and other very peculiar and until recently unsuspected properties of assemblies of matter, and enormous clouds of molecules, some of them "pre-organic" in the interstellar space. Again this is an indication of the strangeness of the universe relative to our conceptions, which were acquired by previous sets of observations and under previous canons of learning and knowledge.

In physics one gets surprises, too, frequently now in the more technological or practical consequences of some physical discoveries. For example, the applications of the conception and development of holograms and their uses are very perplexing at first. Similarly, the new laser techniques are in general very impressive.

The recent discoveries in biology, revolutionary and promising as they are in their introduction of fantastic new vistas of the future change of the mode of life on our earth, have a different epistemological character. I am struck by the "reasonableness" of the arrangements on which life has been shown to be based. The discoveries of the way living matter replicates, everything that followed from Crick's and Watson's models, the nature of the biological code and "tout ce qui s'y rattache," as the French say, show on the contrary a sort of very comprehensible and almost nineteenth-century type of mechanical arrangements which do not require basic physics for the understanding of how they operate. Quantum theory is important for the explanation of the phenomena of the basic molecular reaction, the basis of the arrangement, but the arrangements themselves seem to be quasi-mechanical or almost quasi-engineering in the way they involve the very fabric of the life processes.

One could ask why is this so? Why is it that our understanding of the physical world and perhaps the world of living matter or ourselves or the pattern of our thoughts, does not seem to proceed or accrue continuously? Instead of a logical development of steady growth we observe discrete quantum stages. Is it that the world is really simple in its ineffable structure, but that the apparatus of the nervous system which brings it to consciousness or renders its understanding communicable must, of necessity, be complicated?

RANDOM REFLECTIONS

Is it that the structure of our brain with all its neurons and connections, an admittedly very complicated arrangement, is not best suited to a direct description of the universe? Or perhaps the other way round, reality is on some very complicated objective scale which we do not yet even conceive and we, in our simpleminded way, try to glean it and describe it by simple steps in successive approximations as Descartes prescribed in his *Discours de la Méthode*?

(For a more detailed consideration of the possibilities of a future role of mathematics in biological research, I refer the reader to an article I wrote entitled "Some Ideas and Prospects in Biomathematics." The technical aspect of the article is somewhat beyond the scope of these general remarks, but interested readers may like to look at it.)

In social sciences, a layman like myself feels that there is no theory or deeper knowledge at the present time. Perhaps this is due to my ignorance but I often have the feeling that by just observing the scene or reading, say, *The New York Times*, one can have as much foresight or knowledge in economics as the great experts. I don't think that for the present they have the slightest idea what causes the major economic or socio-political phenomena except for the trivialities everyone should know.

A development whose effect we cannot even estimate and which would have an impact greater by far, I think, than any of the established religions, will be the discovery of the existence of other intelligences in the universe—perhaps thousands of light-years away from the solar system. It is entirely possible that there are waves which have traveled for a long time which we could suddenly decipher. An inkling or a proof of existence without the possibility of communicating two ways would have an overwhelming effect on humanity. This could happen very soon, and it might create panic or, on the contrary, a new type of religion.

We have all read about flying saucers and other unidentified flying objects. Edward U. Condon directed a very thorough study of the subject. Most cases were easily proved to be either illusions, optical or otherwise, or natural atmospheric phenomena, but there remain a few cases of authenticated UFO's which are most puzzling. Take the group of Mount Wilson astronomers who were on a walk and saw a very strange meteoric object, and when they returned to the observatory saw indications of high peaks of radioactivity. There are also a few cases of objects which were simultaneously followed visually, by plane and by radar, which have never been explained.

Fermi used to ask: "Where is everybody? Where are the signs of other life?"

In my opinion, more than anything else it is the new biology which will change the way of life in our world during the next ten or fifteen years. Discoveries which at first seemed rather ordinary have already had more effect on the composition of the world than even big wars: new drugs like penicillin on the one hand, contraceptives on the other, have changed the balance of population.

To illustrate the fast pace of discoveries in biology, I recently heard of two important developments in cancer studies in the space of one week. One is that a Michigan scientist has discovered a virus in the human breast cancer cell. The other is that an experiment which was actually performed in Boulder, where there is a very good electron microscope, resulted in a surprising new technique. Keith Porter and his associates were able to produce cells from which the nuclei could be extracted. These undamaged nuclei can be transferred to other denucleized cells, so in effect it is an exchange of nuclei between cells. For example, a cancerous cell can have its nucleus removed and put into a healthy cell. Then the new cell may become normal. This is most remarkable and shows that some instructions may be coming not from the nucleus, as was believed, but from the cytoplasm.

In the future, new ways to produce or replace food will have much more influence on the shape and mode of human life here on earth than any politico-socio-economic developments in the present sense of the words. All this may be obvious, but sometimes obvious things need to be repeated over and over before they are realized. The world will be so different. I am reminded of a book which was published not so long ago under the title *The Next Million Years.* What lack of imagination it shows!

Gamow's interests, von Neumann's foresight, Banach's and Fermi's work, among others, have all contributed to enlarge the aspect of today's science, and enormously widen the perspectives of physics and mathematics. It is a marvel that so many new vistas and achievements are due to such a fortuitous and fortunate confluence of all the divisions of science.

Selected Bibliography

Index

SELECTED BIBLIOGRAPHY

BOOKS

BY S. M. ULAM

Collection of Mathematical Problems. New York: Interscience, 1960.
Sets, Numbers and Universes. Cambridge, Mass.: The M.I.T. Press, 1974.

EDITED AND TRANSLATED BY S. M. ULAM

The Scottish Book: A Collection of Problems. Translated from a notebook kept at the Scottish Coffee House for the use of the Lwów Section, Polish Mathematical Society. Los Alamos, N. Mex.: Los Alamos Scientific Laboratory, 1957.

BY MARK KAC AND S. M. ULAM

Mathematics and Logic. New York: Praeger, 1968.

ARTICLES

BY S. M. ULAM

"Combinatorial Analysis in Infinite Sets and Some Physical Theories." *Review of Society of Industrial Applied Mathematics,* vol. 6 (1964), pp. 343–355.

"Gamow and Mathematics." In *Gamow Memorial Volume.* Boulder, Colo.: University of Colorado Press, 1972.

"Ideas of Space and Space Time." Rehovoth, Israel: Weizmann Institute, winter 1972–73.

"Infinities." In *The Copernican Volume of the National Academy of Sciences,* Cambridge, Mass.: The M.I.T. Press, 1974.

"John von Neumann, 1903–1957." *Bulletin of American Mathematical Society,* vol. 64, no. 3, pt. 2 (1958), pp. 1–49.

"On the Monte Carlo Method." In *Proceedings,* Symposium on Large-Scale Digital Calculating Machines, September 1949. Cambridge, Mass.: Harvard University Press, 1951.

(307)

SELECTED BIBLIOGRAPHY

"Some Ideas and Prospects in Biomathematics." *Annual Review of Biophysics and Bioengineering*, vol. 1 (1972), pp. 227–292.

BY P. R. STEIN AND S. M. ULAM
"Experiments in Chess on Electronic Computing Machines." *Computers and Automation*, vol. 6, no. 9 (1957), pp. 14–18.

INDEX

Abel, Niels Henrik, 207
AEC, *see* Atomic Energy Commission
Agnew, Harold, 192
Air Force Space Committee, 255, 256
Alembert, Jean d', 272
Alexander, James, 70, 244
Alexandroff, Paul, 46, 66, 68
Alvarez, Luis W., 211
Anderson, Senator Clinton P., 72, 254
Archibald, Raymond C., 100
Archimedes, 104, 274, 290
Argo, Harold and Mary, 153
Aron, Françoise, *see* Ulam, Françoise
Aronszajn, Nachman, 36
astronomy, cosmology, 16 ff, 203 ff, 221, 255, 258, 299
Atkinson, Geoffrey S., 149
atomic bomb, 5, 142, 146, 149 f, 152, 259, 170, 188 f, 209, 211, 221 f
atomic energy, 189, 237; *see also* nuclear energy
Atomic Energy Commission, 193, 210 ff, 221, 224, 228, 231, 237, 240, 243, 253; General Advisory Committee of, 216, 220
Auerbach, Anna, *see* Ulam, Anna
Auerbach, Herman, 42

Auerbach, Karol, 56
automata, 32, 82, 96, 98, 241 f, 285; *see also* electronic computers
Ayres, William L., 107

Bacher, Robert F., 150
Baker, Nicholas, *see* Bohr, Niels
ballistics, 137, 231
Banach, Stefan (1892–1945), 6, 22, 26, 31 ff, 38, 40, 49, 54 f, 65, 107, 180, 203, 245, 293, 303
Bardeen, John, 86
Bell, Eric Temple, 291
Bell, George, 259
Bellman, Richard, 131 f
Bergman, Stefan, 119
Bernouilli law of large numbers, 73
Bernstein, Dorothy, 131
Bernstein, Felix, 27
Besicovitch, Abraham, 59 f
Bethe, Hans Albrecht, 143, 147, 149 f, 152, 156, 162, 172, 191 ff, 216, 260, 263
Beyer, William, 268
Bigelow, Arthur, 229
biology, 203 ff, 242, 259 f, 300 ff
Birkhoff, Garrett, 66 f, 84, 89 f, 90, 92, 120
Birkhoff, George David (1884–1944), 6, 74 ff, 82, 85, 89, 90,

INDEX

Doyle, Sir Arthur Conan, 59, 145
Dykstra, Clarence A., 126
Dyson, Freeman, 256

Eddington, Sir Arthur S. (1882–1944), 60, 62
Eilenberg, Samuel, 66, 68, 134
Einerson, Ivar, 86
Einstein, Albert (1879–1955), 4, 6, 18, 47, 62, 73, 81, 108, 112, 157, 167, 184, 205, 223, 272, 294
Eisenhower, President Dwight D., 243, 251, 253, 255
electronic computers, 4, 13, 96, 137, 155, 181 f, 197 f, 200 f, 212 f, 215 f, 218, 225 f, 229, 241, 244, 257, 264, 284 ff; *see also* automata
Eliot, Charles W., 87
Elliott, Josephine, 215
Erdös, Paul, 53, 54, 60, 131, 134 ff, 178 f, 294
Ernst, Marcin, 16 f
Estreicher, Karol, 30
Euler, Leonhard (1707–1783), 20, 104
Evans, Cerda and Foster, 213
Evans, Herbert, 130
Everett, C. J., 31, 104, 124, 129, 131, 160, 195, 213 ff, 219 f, 225, 253
evolution, 205

Feinsinger, Nathan P., 125
Fejer, Leopold (1880–1959), 110
Fermat, Pierre (1601–1665), 48, 278, 291
Fermi, Enrico (1901–1954), 6, 152, 157, 161 ff, 172, 184, 191, 193, 215, 217 ff, 223, 225 ff, 233 ff, 243, 245, 257, 303
Fermi, Laura, 234, 236 f, 282

Feynman, Richard, 77, 150, 152 f, 156 ff, 162, 169, 221, 252, 297
Fisk, James B., 86, 193
fission bomb, *see* atomic bomb
Flanders, Donald A., 169
Ford, Kenneth, 220
Ford, Colonel Vince, 255
Foucault pendulum, 230
Fowler, William A., 257
Fraenkel, Stanley, 153, 184 f, 213
France, Anatole, 23, 123, 224
Freud, Sigmund, 18, 22
Frisch, David and Rose, 144
Frisch, Otto, 6, 162
Froman, Darol K., 219

Galois, Evariste (1811–1832), 207, 277
Gamow, Barbara, 205
Gamow, George (1904–1968), 100, 151, 202 ff, 211 f, 231 f, 266 f, 270, 294, 303
Gamow, Rho, 233
Gardner Space Committee, 256
Gardner, Trevor, 255 ff
Gaulle, General Charles de, 257
Gauss, Karl Friedrich (1777–1855), 231, 289
Getting, Ivan, 86
Goad, Walter, 259
Gödel, Kurt, 55, 76, 80, 283
Goldstein, Max, 218
Gorman, Elizabeth S., 244
Gould, Gordon, 259
Graustein, William Caspar (1888–1941), 75, 89
Greenstein, Jesse L., 257
Griggs, David, 257
Grossman, M., 57
Guerlach, Henry, 86
Gurney, Herbert W., 267
Guttierrez, J. J., 168

(311)

INDEX

INDEX

Ulam, Andrzej (cousin), 72, 115
Ulam, Anna (mother), 9, 15, 113
Ulam, Claire (daughter), 161, 174, 230, 235, 239, 244
Ulam, Françoise (Aron), 44, 125, 133, 145, 161, 174 ff, 178, 189, 190, 215, 230, 239, 244, 260, 262, 269, 270, 272
Ulam, Jozef (father), 9 ff, 18 ff, 113 f
Ulam, Michael, 56 f, 108
Ulam, S. M.
  birth, 9
  brain illness, 174 ff
  in Cambridge, England, 58 ff
  Colorado, University of, 266 ff
  doctor's degree, 47
  Los Alamos, 141–171, 188–208, 209–224, 249–265
  marriage, 125
  mathematical curiosity, early, 20 ff
  "self-portrait," 271 f
  U.S. citizenship, 137 f
  World War II, outbreak of, 116
Ulam, Szymon (uncle), 16, 114
University of Colorado, 91, 161, 260
University of Southern California (USC), 173 f, 179, 186
University of Wisconsin, 62, 120, 123, 125

Vassilief, Alexander, 124
Veblen, Oswald, 73 f, 115, 283
Veblen, Thorstein, 74
Verne, Jules, 5, 23, 255, 229
Vienna, 10 f, 56 f, 261
Virgil, 111

Vitali, G., 92
Voltaire, 105, 262
Von Neumann, John, see Neumann, John von

Walden, William, 201
Wang, Hao, 285
Watson, James D., 241, 260, 300
Wavre, Roland, 45 f
Weierstrass, Karl, 55, 282
Weil, André, 133
Weisskopf, Victor F., 127, 152, 162, 168 f, 172, 191, 260 ff, 270
Weizsäcker, Carl Friedrich von, 216
Wells, H. G., 5, 23
Wells, Mark, 201
Weyl, Hermann (1885–1955), 30, 72, 121, 291
Wheeler, John, 215, 220 f
Whitehead, Alfred North (1861–1947), 29, 86, 118 f, 138, 285
Whitehead, Mrs. A. N., 30, 118
Whitney, Hassler, 91
Whyburn, Gordon T., 107
Wiener, Norbert (1894–1964), 45, 91 ff, 109, 137, 160, 294
Wiesner, Jerome, 253 ff, 257
Wigner, Eugene, 112, 124, 160, 277
Woodward, Robert B., 86
Wundheiler, Alexander, 119

Yang, Chen Ning, 294
Young, Lawrence C., 62

Zabuski, Norman, J., 228
Zawirski, Zygmund, 21, 23, 65
Zygmund, Antoni, 38, 93, 96, 270